BREAKPOINT

JEREMY B. C. JACKSON AND STEVE CHAPPLE

Breakpoint

*Reckoning with America's
Environmental Crises*

Yale UNIVERSITY PRESS

NEW HAVEN AND LONDON

Published with assistance from Furthermore: a program of
the J. M. Kaplan Fund.

Furthermore:
a program of the J.M. Kaplan Fund

Yale University Press books may be purchased in quantity
for educational, business, or promotional use. For
information, please e-mail sales.press@yale.edu (U.S. office)
or sales@yaleup.co.uk (U.K. office).

Set in Janson type by Westchester Publishing Services,
Danbury, Connecticut.
Printed in the United States of America.

Library of Congress Control Number: 2017958069

ISBN 978-0-300-17939-2 (hardcover : alk. paper)

A catalogue record for this book is available from the
British Library.

This paper meets the requirements of
ANSI/NISO Z39.48-1992 (Permanence of Paper).

10 9 8 7 6 5 4 3 2 1

To Cody, Jack, Julia, Rebecca, and Stephen, and
to your future in a healthier and safer world

CONTENTS

The ideas for this book grew out of the shockwaves following the *Deepwater Horizon* oil spill. We had met two years before, when Steve, a veteran journalist and adventurer with strong environmental leanings, enrolled in an interdisciplinary master's program for aspiring ocean leaders taught by Jeremy at the Scripps Institution of Oceanography. While discussing the challenges of communicating the environmental issues presented in class, we discovered that we shared similar perspectives on the world and soon became friends.

Right after the *Deepwater Horizon* disaster in 2010, Steve traveled to the Gulf to report on the ensuing chaos, while Jeremy, who had led the Smithsonian Institution's investigation of the consequences of another hugely destructive oil spill in Panama in 1986, fielded questions from the media about the implications of that earlier study for the Gulf of Mexico. Before long, Steve began working on the idea for a book about the oil spill and its consequences and came to Jeremy for advice. But Jeremy, an ecologist and geologist who specializes in understanding the impacts of human activities on the planet, knew that the spill was just the most recent disaster to afflict a region that has long been disrupted by human mismanagement and interference. Horrible as it was, the oil spill was a spectacular distraction from a far greater ecocatastrophe that has been building toward a breakpoint for over a century.

Serious and lasting environmental damage to the Mississippi Delta region began in the early 1900s, when the far northern headwaters of the mighty Mississippi and Missouri rivers and their tributaries in Minnesota and the Dakotas were channelized, diverted, and dammed so extensively that the river was starved of its nourishing sediment. Steve's mother was from the Dakotas, and he had grown up in Montana and knew about the damming of the rivers firsthand.

A few decades later, countless tons of artificial fertilizers and pesticides derived from petroleum began to be sprayed, dumped, and injected onto and into the vast fields of the Corn Belt to sustain a remarkable agricultural revolution that was widely heralded as the miracle that would feed the world. But runoff of the residual nitrogen, phosphorus, and poisons of the new industrial agriculture began to course downstream. This toxic slurry, diverted by dams and levees, spread across the Mississippi Delta, weakening the very fabric of the vast coastal marshes that had for millennia protected the shoreline from storms.

Ultimately reaching the Gulf of Mexico, the wasted nutrients caused an explosion of microscopic algae whose rotting remains sucked up all the oxygen in some of the coastal waters off of Louisiana and east Texas, forming a "dead zone" as big as New Jersey that is devoid of fish and shellfish. Meanwhile, decades of dredging to construct thousands of miles of pipelines for oil and gas production, as well as widened shipping channels, hastened the "death by a thousand cuts" of the marshes that, at least before the horror of Katrina, had protected New Orleans from hurricanes.

As if all this weren't enough, climate change caused by the same oil and gas industry that manufactures all the fertilizers and pesticides and has sliced up the Mississippi Delta is making all these problems worse. Alternating periods of extreme drought and torrential rains are strongly affecting corn and soy production, while the rise in sea level due to global warming threatens to drown large swaths of the East and Gulf coasts from Maine to Texas.

These problems are the byproduct of two of the most important and successful components of American prosperity: the petrochemical industry along the Gulf, and industrial agricultural production in the upper Mississippi Heartland. "Big Agriculture" produces more than a third of the corn and soy grown worldwide, as well as the grains that are consumed in countless forms—from the meat, poultry, and processed food we eat to the ethanol in our gas tanks. But this prosperity seems increasingly hollow as the risks to the safety of our homes and health skyrocket.

Why aren't we doing anything about this? One reason (or excuse) is that environmental threats are frustratingly abstract. Our eyes tend to glaze over as scientists attempt to explain the causes and consequences of a 1-, 2-, or 3-degree Fahrenheit rise in temperature over the next ten, fifty, or hundred years; and our appallingly inadequate educational programs in basic science do little to counter voters' distrust of things they don't understand. But it's not so abstract if you're a farmer watching your crops being destroyed again and again by increasingly persistent droughts or if you were unlucky enough to have been in New Orleans or New York City during Hurricanes Katrina or Sandy. Or in Houston, Florida, and Puerto Rico during the horrific hurricane season of 2017.

So we set out to explore the Heartland and the Gulf firsthand to talk to people, listen to their stories, and see for ourselves what was going on. We began at Lake Itasca, Minnesota, the headwaters of the Mississippi where the river trickles forth in a stream we could step over, then worked our way south through the corn and soy shag of Iowa, past the gigantic systems of locks and levees, and through Louisiana's Cancer Alley where massive refineries vomit out chemical waste. Then on to Katrina-besmacked New Orleans and the once-proud Bird's Foot Delta with its bountiful marshes, shrimp, oysters, wildlife, and natural gas that are sinking into the ocean at the shocking rate of a football field every day. It was a journey that few Americans living on either coast ever make, full of fun, adventure, great people, food, and music, sublime natural beauty, and nature unnaturally scarred.

Everywhere we visited seemed on the verge of a breakpoint. Midwestern corn was reaping enormous profits—although these have since collapsed—but the challenges of farming were increasing due to everything from pesticide-resistant superweeds to dramatic increases in extreme weather events that were destroying crops and causing catastrophic soil erosion and loss. Pollution from excess nutrients was poisoning the region's drinking water and creating dead zones in Lake Erie even more threatening than the dead zone in the Gulf of Mexico. Oil and gas were booming despite a recent drop in gas prices at the pump, and natural gas was winning the competition against coal as the major source of cleaner energy for generating electricity. But the entire Mississippi Delta seemed on the verge of disappearing beneath our feet and with it the great marshes' centuries-old buffering protection for New Orleans and the entire Louisiana coast.

Back home we reflected on how the fate of places where we had stopped and people whom we had met hold great meaning for all Americans, more than most of us know. In our minds, our journey became a metaphor for the range and scope of the environmental problems we all face, and of how the things we do in one place affect the wellbeing of faraway places and peoples we have never met—not just in the Heartland or the Delta, but everywhere that human actions are pushing the environment beyond its limits. The list is daunting. Severe long-term drought in the Southwest. Soil erosion, superweeds, and poisoned groundwater in the Heartland. Sea level rise, land subsidence, and stronger hurricanes along the Gulf and East coasts. The obesity crisis fueled by our appalling diet. Change creeps up so slowly that we hardly notice. A bad hurricane here, an exceptional flood there, the drinking-water crisis in Toledo, Ohio; all shrugged off as exceptional events while people soldier on and adapt as best they can. That is, until suddenly everything unravels and one's home, one's livelihood, will never be the same again.

And then in the summer of our final writing the unraveling began. The fires in the American West smoldered here and there at first before bursting forth in massively destructive firestorms fueled by a too wet winter and a long hot summer. At the same time, first one, then two, then three massive hurricanes crashed into the places to the south of where we had been, and Puerto Rico, too. Harvey, Irma, and Maria are names that will not soon be forgotten as costs of their combined destruction progressively rise to something like half a trillion dollars. Human suffering is immense, especially in Puerto Rico, where the magnitude of misery, social dislocation, and emigration could well exceed that of Katrina.

None of this should have been a surprise. We understand fundamentally how warmer global temperatures caused by greenhouse gas emissions from burning fossil fuels are increasing the occurrence of extreme weather events including tornadoes, droughts, and storms. No informed person today can deny that hotter oceans are fueling more catastrophically powerful hurricanes. Scientists had predicted that 2017 would be a bad hurricane year and they were right. This is the first time in recorded memory that three catastrophic Atlantic hurricanes have hit the United States—the previous record was just one.

We had predicted much of this in dry prose, and Jeremy had been lecturing about it for years in places like the Naval War College, but even we were stunned at the rapid-fire intensity of what hit America in the summer of 2017. Science had foretold all of this without controversy, at least to scientists. But it was a burning epiphany to many smart and busy people who hadn't given "extreme weather" much thought, and it was perhaps a soggy blow across the head to those who had only months before called climate change a Chinese hoax. Or maybe not? Much money is to be made by denying for as long as possible the new normal, though, in our view, much more money will be made, more jobs and prosperity, and better health, by coming to terms with what we all know to be true. But we run ahead of ourselves now in the sober spring of 2018, wondering what this summer, and summers to come, will bring.

We need as a people to come to grips with all these real and severe challenges today before they spiral even further out of control. And we need real leadership. Responsible, caring, and courageous leadership that goes beyond the corruption and deliberate misinformation promulgated by the special interests that dominate American politics and the Republican Party's war on science today. Leaders with the courage to adopt measures for the long-term wellbeing of all of us rather than the short-term profits of the privileged few. It will not be easy. It's too late for quick fixes and Band-Aids. But there are innumerable tractable and practical solutions that really work or are being newly brought on line that offer great social and economic opportunities as well as some inevitable loss.

Americans have faced and overcome great challenges before and we can do it again. We hope that this book will contribute to the new thinking that will lead the way.

ACKNOWLEDGMENTS

Our deep and sincere thanks to all the wise and wonderful people we met along the way who took the time to explain what they do and to reflect on the future of the Heartland, coasts, and world beyond. Some of them helped us out on repeated visits. John Weber set the stage for our explorations of Iowa agriculture and patiently explained the workings of a state-of-the-art GMO industrial farm. Gene Turner and Foster Creppel made it all happen on the Mississippi Delta. Gene traveled down from Baton Rouge with his boat and vast experience to tutor us on the ecology of marshes. He also organized and financed the overflight that was so critical to our understanding of how deeply wounded the entire region is. Foster, who seemed to know absolutely everyone around the Delta, set us up to meet John Tesvich and Erik Hansen and regaled us with his hospitality and strong opinions about the future of the Delta and what to do to save it. John Lopez spent an entire day in the field showing us how he studies the slow buildup of sediment in the Bohemia Spillway. Kristin Rechberger patiently explained the guiding principles of the green economy and provided much of the data on renewable energy production and subsidies in the epilogue. Rob Dunbar, Shirley Laska, William Nordhaus, Jeffrey O'Hara, and Nancy Rabalais provided additional deep insight. Karen Alexander read the entire manuscript twice and provided extensive comments and criticism. She also transcribed the majority of our interviews and edited the endnotes. Joey Lecky skillfully produced the maps.

Special thanks to our wise and more than patient editor Jean Thomson Black at Yale University Press for guiding us through the process of organizing our somewhat disconnected thoughts into a coherent book, and to Michael Deneen and Margaret Otzel for their technical editing advice and support.

The Scripps Institution of Oceanography and Smithsonian National Museum of Natural History provided welcoming refuges to write and much institutional support. Penny Dockery and Laura Schaefer helped out with travel and logistics. The Ritter Chair of the Scripps Institution of Oceanography supported most of our travel. We are also grateful for the generous support of our research and writing from Furthermore: a program of the J. M. Kaplan Fund, the Ocean Foundation, the New Orleans Foundation, Charles J. Smith, Sheldon Engelhorn, Fritzi Cohen of the wonderful Tabard Inn in Washington, D.C., Mike Condon, B. Greg and Betsy F. Mitchell, Jack Nooren, and Susan and Max Guinn.

Steve thanks his occasional traveling companion, the hardy Jack Chapple, and not least, the forbearance and wisdom of his wife, Ines Salgado Chapple, who put up with late nights and distant travel. Jeremy thanks his wife, Nancy Knowlton, for cheerfully putting up with the two of us during our intense writing sessions in Maine and for her wisdom, optimism, and love.

BREAKPOINT

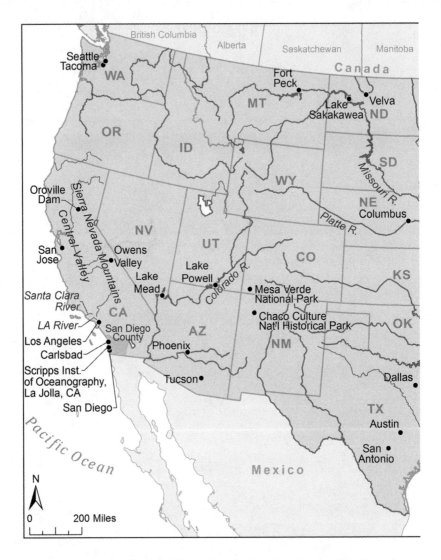

The lower forty-eight United States showing locations discussed in this book.

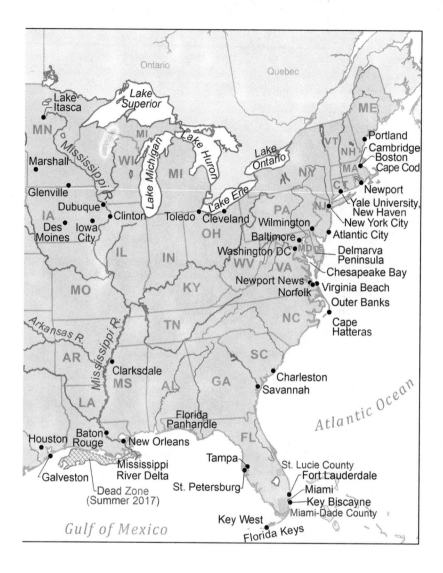

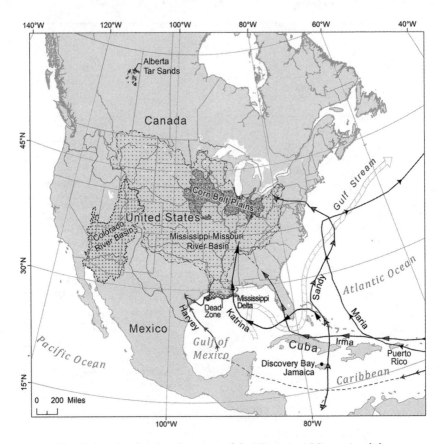

North America showing the extent of the Mississippi-Missouri and the Colorado river drainages, and the areas of the Corn Belt and dead zone in the Gulf of Mexico. Also indicated are the paths of Hurricanes Katrina, Sandy, Harvey, Irma, and Maria; the course of the Gulf Stream; and locations beyond the lower forty-eight states that are referred to in this book.

PART I

Heartland

John Weber, sixty-three, climbs into the driver's seat of his quarter-million-dollar tractor, ten feet up from good Iowa dirt, and surveys the fields before him. He is happy and prosperous, and what he sees has made America happy and prosperous, too.

In the beginning of that summer in 2011, the economic future for agriculture looked so bright that some Iowa farmers bought and plowed up golf courses. Others moved old homesteads to turn house lots into productive fields. Still others put acreage that had been set aside for conservation back into production once their Conservation Reserve Program (CRP) contracts were up. Hilly land, marginal land, and conservation land were all put to use to plant more corn and soy.[1]

But by mid-July, a record-setting twenty-one straight days of temperatures over 90 degrees had dried out the soil. Weber had started the season with adequate moisture in the ground, what Iowa farmers call "a full tank." Other farmers were not so lucky. "Any areas that didn't have rain or good subsoil wilted down to paper. Here we were fortunate to get rain," he told us, "and the subsoil was good. We're kind of taking a beating on beans. They are starting to turn. They are dropping their leaves too rapidly. We had high hopes for Hurricane Isaac but only southeast Iowa got any of it." And the soybeans were "like beads of glass, hard as bone, dust everywhere."[2]

A pinpoint wind event swept across the six counties surrounding Weber's farm the day before our first visit in August. Gusts were clocked at 129 mph and were sustained for twenty-eight minutes. Weber lost ten grain bins. Hundred-year-old oaks blew down. Weber, president of the National Pork Producers' Council, was away at a meeting of hog farmers in the Wisconsin Dells. At 4:00 in the morning his wife, Cathy, called. "You better get back here," she said. He drove six hours straight, but by the time he arrived his son, Brian, and hired hands had already started to clean up, beginning "to put forty years of work back together in three months. Friends and neighbors shared equipment and cleaned the debris," Weber recalled.

The soybeans were low enough to the ground not to be hurt, and most of the corn was okay, too, though twisted. "It'll be heck to drive the harvester through, but we'll get most of it," Weber predicted.

On our first walk around Weber's farm, we passed through the barn. The machines were new and gigantic and had been purchased outright. Tires on the spreaders were taller than we were, so the machines could slide along above the tassels. Adjustable booms could be attached to extend as far as sixty feet on both sides. The spreader, the tiling machine, and the harvester are all guided by global positioning, that is, they are driven from space.

In the government-regulated world of American production agriculture, there are winners and losers, and John Weber is a winner. A fourth-generation Iowa farmer who lost a farm during President Jimmy Carter's agricultural boycott of the Soviet Union, he recalls that the loss taught him to never forget that "this is a high-risk business, and the point is to reduce risk." This year's drought-damaged harvest was even more valuable than the full harvest the year before. From May to October, the price of corn nearly doubled, reaching a futures price of $8.20/bushel before settling lower.[3]

With Brian and Cathy, his wife of forty-five years, John Weber raises corn and soy on their 2,500 acres (he owns or leases several other Iowa farms as well). The family contracts to grow approximately 750 acres of seed corn, which is raised for the Pioneer Seed Company next door. Brian Weber is Pioneer's regional seed representative. This seed corn is mostly GMO (indicating that it's a genetically modified organism), with multiple hybridized traits that allow it to be used in concert with Roundup herbicide, which is applied to the tops of the plant. Other GMO seeds, according to Weber, work to kill rootworms or help the plant to resist drought or simply produce a higher yield. The Webers also "finish" some fourteen thousand hogs, buying shoats—feeder pigs weighing about fifty pounds—and raising them up to a market weight of 250 to 300 pounds, depending on the breed. At the time of our visit, Weber's pigs were under contract to Cargill, a privately owned agribusiness corporation with $135 billion in annual sales, but in 2015 Cargill agreed to sell its pork division to JBS S.A., the Brazilian company that is the leading processor of animal protein in the world, with over 200,000 employees.[4]

Iowa is the top hog-producing state in the United States. It is also first in corn production and second in soybean production. Iowa produces three times more corn than Mexico, where corn or maize originated, and about 11 percent of the global crop.[5]

Yet these days, little of the corn raised in Iowa or the other Heartland states (in descending order of corn production, Illinois, Nebraska, Minnesota, Indiana, South Dakota, Kansas, Wisconsin, Ohio, and Missouri) goes to feed people, at least not directly. In 2015, 41 percent of U.S. corn production (five billion bushels, or 290 billion pounds) was converted to ethanol for cars and trucks; 26 percent, including a byproduct of ethanol distillation called dried distiller's grains, became feed for hogs, cattle, and chickens (in that order); about 18 percent was transformed into industrial and residual foods including high-fructose corn syrup to sweeten soft drinks and processed foods; and 15 percent was exported. Less than 1 percent was the sweet corn that we eat eagerly at summer barbecues.[6]

Iowa, perhaps the most iconic agricultural state, produces little real food. To say, then, that America's Heartland feeds the world would be strangely misleading. More accurately these days, it helps to power the world, generate its industrial sweeteners, and feed its hogs. The region's most important role by far is to supply energy. Biofuel production has risen eightfold since 2000. And the pace is accelerating—from 6.5 billion gallons in 2007 to nearly 15 billion gallons in 2015.[7] The Webers believe this shift will change American agriculture forever. In what has been called the "Food vs. Fuel" debate, a debate far more nuanced than either side admits, fuel now has the upper hand.[8]

This trajectory is good for the Webers, who are well acquainted with the ups and downs of commodity markets and appreciate the high demand. It's also good for America's balance of payments as corn and soy are shipped to China, Canada, Mexico, and other countries.[9] But it's not so good for the ecology of Iowa or for states downriver, especially Louisiana, or for the Gulf of Mexico. This is because Iowa's soil covers a vast network of buried pipes perforated with billions of drain holes. The pipes are called tiles, since they were once made of ceramic pottery. Tile drainage of fields has been used in one form or another since at least the time of the Romans more than two thousand years ago. The tiles prevent plant roots from getting soggy: corn, especially, enjoys well-drained soils. As in the other Heartland states, this plumbing system channels excess water—and what's in the water—away from fields, into streams, and eventually down the Mississippi River. Iowa spans 62,000 square miles, of which about one quarter is tiled by roughly half a million miles of tiles—the equivalent laid end to end of twenty times around the earth and an engineering feat that dwarfs the transcontinental railroads, the interstate highway systems, and the Great Wall of China in sheer distance covered. In short, much of Iowa functions like a giant potted plant with drain holes.[10]

As we drove back to the Weber's farm the week before Thanksgiving, the coffee-brown fields were dotted with tractors pulling large white cylinders of anhydrous ammonia, which is a byproduct of natural gas refined in chemical plants along the Mississippi north of New Orleans (a stretch of the river known as "Cancer Alley") as well as in several plants in Iowa. Over the winter, the ammonia converts to nitrate fertilizer in the soil. Ammonia is best applied when the ground temperature is 50 degrees Fahrenheit or colder. Up to sixteen "knives" pulled behind giant tractors slice the soil to instantly cover the gas as it is pumped into the dirt.[11] By spring, the ammonia gas will have magically transformed into a form plants can use, a process that has increased U.S. corn yields fourfold since 1950.

Tiled water from the Weber farm, in Iowa's northeast quadrant, trickles to the edge of his fields, tumbles into several unnamed ditches, and flows to Wolf Creek. Wolf Creek empties into the Cedar River, which in turn empties into the Iowa River just before it slides into the Mississippi near Muscatine. Water from the Weber farm then rolls another 724 miles to New Orleans and, crossing the Bird's Foot Delta, rushes far out into the Gulf of Mexico. The Heartland's tile drainage system disgorges about 1.5 million tons of excess nitrate and phosphorus fertilizer into the deep water of the Gulf. To understand what's happened to the Gulf, then, one must first understand what's going on in the Heartland.[12]

"I love driving my tractor these days because I don't have to do any driving," jokes Weber. He closely monitors the controls from inside his air-conditioned cockpit, like a middle-aged jet pilot. The newest high-tech development, Variable Rate Technology (VRT for short), allows him to program his "inputs," everything from phosphorus and nitrate fertilizers to pesticides, before he even climbs into the tractor. Soil samples are taken from three-acre plots on a grid and tested for input needs. The results are then programmed into the on-board computer, which automatically delivers what's needed to each three-acre plot as the satellite guides the tractor. If the 120-foot spreaders go over ground needing fewer inputs, the nozzles shut off.

"You hear them go, click, click, click, click, click and then they turn back on again. These new technologies are huge to us," explains Weber, who is also quick to point out that this pinpoint delivery system means fewer nitrates and herbicides and less phosphorus go into the ground. He then makes a pitch for GMOs, which have been used in Iowa for over a decade. "Every farmer seriously weighs the use of GMO because it is expensive," he explains, but "if you have warm winds, that can, for instance, bring an insect invasion, moths and corn borers," then GMO seeds, to Weber, are one preventative. "None of this would be possible without GMOs, none of it. Some people don't understand. Corn that has been modified requires less herbicide, less pesticide, and so in the long run, it's safer. It's certainly safer for the applicators, more so than the way my dad and I used to apply organophosphates to the fields."

Our conversation turned to commercial uses of corn and soy. High-fructose corn syrup is produced in plants like the gargantuan

Archer Daniels Midland (ADM) Cedar Rapids facility less than forty miles away. Weber knows the knocking that high-fructose corn syrup and processed food get from critics on both coasts—including former New York City mayor Michael Bloomberg, who tried to ban super-size soft drinks because of their adverse health effects. Weber knows about the diabetes and heart disease.[13] He's also aware of the import tariffs on cane sugar arriving from Brazil and the Caribbean. "You can't tell me there is any difference chemically between high-fructose corn syrup and cane sugar!" Weber protests, and he is basically right.

Weber is remarkably fit, over six feet tall, thin and well-muscled. "You just got to take care with your diet." He points out that his most valuable pigs are the ones with the most fat. They sell for the highest prices to the Japanese, who value taste above all else. He holds out the palm of his hand and tips it, just so. "That loin steak's got to be firm, bend just right, and the fat must be white, not yellow, and that's determined by what the pig eats. The time it takes for one of my special pigs to be killed and processed, the meat shipped across the country in refrigerated trucks and shipped to Tokyo to a special market—and I've been there, it's something else—is seventeen days."

From hoof to table? "Hoof to table."

Weber understands many things and he worries about many issues. A best-practices farmer, he cares about the future of American agriculture. But production agriculture for corn and soybeans, hogs, and ethanol is what Iowa has become. It's the system. Weber is aware of the Mississippi's dead zone, "that it could possibly be caused by" applying too much nitrate to Iowan and Midwestern fields. "That's something we need to address up here. I do not want to blame the dead zone on nitrogen alone rather than on phosphates, too. We are reducing the amount of nitrates. Ag is not the only source. There are a lot of major communities. What about storm water discharge? It's personally important to me to leave the soil in better condition than when I started farming. We need more monitoring, and we all need to work together."

Problems are there to be solved, "is all," in his can-do worldview. "If you install a pattern system of tiling," for instance, "you can close them in periods of drought. If the next ten days look dry, we can drive by and close the valves."

Weber's been to Russia and China. China's pork industry is very competitive, but China consumes all the pork they produce, he points out. In fact, China bought Smithfield Farms, the largest producer of pork in the world. He points out a few things about China. They purchase $24 billion of U.S. agricultural products each year—$14 billion in soybeans alone—which makes China the largest buyer. In fact, China is Iowa's fastest growing trading partner, buying about $1 billion in farm products in 2015.[14] "They take food security far more importantly than we do in this country. We just go to the grocery store. This is not the case in other nations. The number of people in China creates a huge demand. It's Brazil that will be our competitor."

Next to a row of forty-foot-tall, corrugated steel silos that store "the beans," we pass by one of Weber's hog barns. The hogs, a white crossbreed, are intelligent-looking creatures. They are kept clean, and in hot weather their bodies are sprayed with mist. The barns' slatted floors allow hog manure to fall through for collection below. The manure is then pumped out and recycled to fertilize the corn, cutting down on the need for—and cost of—anhydrous ammonia. If he had more of it, Weber says, he could use the methane gas to power his vehicles. The problem is how to collect enough to make it cost-effective. "Ten farmers would not be enough. Different with cattle and dairy," he says, citing some very large farms in Indiana.

Through the open walls, the green cornfields are evident. We step into a double-cab pickup and drive to a field with long rows of seed corn, each row meticulously labeled with numbers and codes. Soon this corn will be carefully picked and trucked next door to the DuPont Pioneer plant, where it will be sorted, tested, and bagged. In nine months, much of it will be growing throughout Iowa.

We visited again after the seed corn had been harvested. Looking down a long drainage ditch at the border of his farm, Weber pointed:

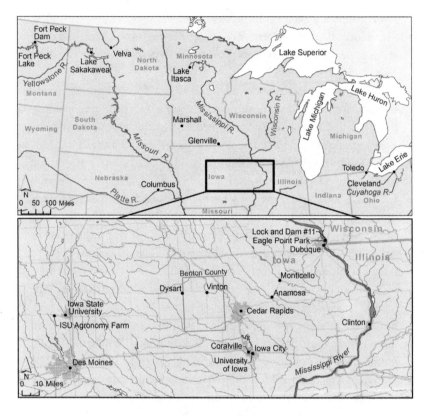

Top: Northern Midwestern and Rocky Mountain states showing
the geographic context of the Heartland's river systems.
Bottom: Detail showing locations visited in Iowa.

"See that yellow flag? That's a tile outlet. The water is crystal clear. Of course, nitrates are the main cause of concern, phosphorus second. You can't see nitrates. It's not like silt. It's not muddy."

Nitrates in Iowa's drinking water are very much a concern, as we shall explore in depth in Chapter 4. The city of Des Moines spends as much as $7,000 a day to filter nitrates out of municipal water. Above certain concentrations, nitrates can cause blue-baby syndrome and have been linked to certain adult cancers.[15] "Des Moines filters it, but then they just release it back in the river, downstream," Weber shrugs. "I ask myself, if the nitrates are too high, what practices can I implement on my farm to get them down? Cover crops, saturated buffer zones for grasses to pull out some of the nitrate—how can less be applied in the fall? Iowa State agronomists have been monitoring with me for the last three years to understand how to get them down. The biggest thing is how and when we put them on in the fall."

Weber alternates soy with corn on most of his fields. Corn residue left in the fields after harvest lessens the erosion he would get from planting continuous soy, but soy residue left on the fields leaves more available nitrogen in the soil. Corn needs the nitrates, and this cropping system lessens the amount of fertilizer it requires.[16] But rotating soy with corn requires more work and more fuel. For this reason, many farmers do "corn-on-corn," following one crop of corn with another and another and another. Corn-on-corn depletes soil nutrients, which are replaced by injecting anhydrous ammonia and applying other chemical fertilizers. The long-term effects on the soil and environment of this artificial "juicing" strategy are unclear. But with the price of corn so high and no penalties for the cleanup of municipal water supplies, of water flowing downriver, or of the Gulf of Mexico, people do-the-math-and-follow-the-money. "End of the day, it's the consumer, the price of the commodity, that determines what you do."

He points to the edge of the long field, under which the pipe is buried. A sixty-foot covered grass border separates the field from the ditch. It filters rainwater and snowmelt, and keeps nitrates from washing into the ditch during "weather events." The natural barrier also holds good dirt on the fields. "I don't know a farmer who doesn't think about his dirt," says Weber, "but some don't see the big picture, the people between the farm and the Gulf of Mexico or even Des Moines. Farmers are conscientious about their farms and their families, but they are busy and active and so they don't always hear about the problems that could be a result of something they are doing. That can be frustrating to me. You have to be aware of how the food you produce also impacts those same people that eat it."

Conservationists feel, and Weber agrees, that if taxpayer money subsidizes farmers, then the public should get a long-lasting benefit for its investment: the preservation of the Heartland's precious soil bank. The system that was developed after the climate catastrophe known as the Dustbowl has by all accounts saved a lot of topsoil. "Farmers have seen significant benefits from cover crops," says Weber, and he lauds Pioneer as "a huge supporter of the Iowa nutrient reduction strategy, with a goal of some 45 percent reduction by 2022, with an incentive program. They will cost share on cover crop seed, and if you want strip tilling, they will give you some money to get you started."

"If farmers like yourself want to keep the best soil, who wants to let it go?" we ask Weber.

"Well, there's arguments in the farm community," he smiles, his eyes twinkling a bit, as he shifts gears on the double-cab pickup to bring us back on the pavement. "Some farmers just want all land to be producing beans and corn. It's 24/7 around here."

"There are farmers that don't want to put CRP [land in the national Conservation Reserve Program] along their streams. There are farmers that are aggressive and grow fence-line to fence-line, but I am seeing so much change in the new generation of farmers who are very concerned in making changes and doing a better job and not spending money that is going down the river. They are educated and understand the chemistry, and also the GMO. As I close out my career, I am in the transition from the go-go years of the 1970s, which created a lot of problems even though we initiated great technology and got high production. You never stop learning. I love to work on the farm. How can I do it better or differently? That's what has always been exciting about farming to me. Try some new technologies and if you get better results, change. It's important to me to keep things heading in the right direction. Represent Midwest agriculture fairly, is all I ask."

Zooming east across Iowa on Interstate 80, the entire state seems like one gigantic cornfield, mile after mile, forever. We fight to stay awake. Oddly, too, while passing through these nonstop gigantic farms, a person could get mighty hungry looking for something to eat. Even on side roads, there were no vegetable stands like the ones all over our home states of California and Maine—and it takes more skill and gumption to grow food for a living on Maine's rocky ground than in Iowa's rich topsoil.

It wasn't always so. Fifty years ago, Iowa produced much of the food that Midwesterners ate. At the same time, farmers in the "Garden State" (a nickname still on New Jersey's license plates) fed all sorts of vegetables, fruits, and nuts to everybody from Washington, D.C., to Boston. Now New Jersey is mostly suburbs, and very few Iowans grow food for people anymore. You can't get as rich growing food as growing fuel.[1]

According to the U.S. Department of Agriculture's most recent assessment, Iowa ranked second among states for the total cash value of its agricultural products, which brought in a walloping $31 billion. Only California surpassed it with $43 billion in sales. California is nearly three times larger than Iowa, but Iowa has 15 percent more active agricultural land than California does. Apart from dominating U.S. agricultural markets, the two states could not be more different. That's because California grows an abundant variety of food for people to eat, whereas Iowa grows industrial quantities of two crops, corn and soy. Most of the food crops that Iowa grows reaches dinner tables—and engines—as something else.[2]

In 2012, California was number one in total production of vegetables, fruits, and nuts, which were valued at $24 billion and harvested from over 4.3 million acres.[3] In contrast, forty-second-ranked Iowa sold about $100 million in fruits and vegetables grown on about ten thousand acres. Just as in New York City, most of the fruits and vegetables we saw in Iowa supermarkets came from somewhere else—most likely California, Florida, Mexico, Chile, or Peru. There isn't much room for growing food in Iowa with all that industrial corn and soy.

Iowa was the number one producer in 2012 of grains, oilseeds, and dried beans and peas, with a staggering 2.4 billion bushels of corn and 462 million bushels of soy valued at more than $17 billion (ten times more than nineteenth-ranked California). Most of the soy goes into edible oil and animal feed, with increasing amounts converted to biodiesel.[4] But as we learned in Chapter 1, most Iowa corn never makes it to a supermarket, at least not directly.

Forty miles south of Weber's farm, we approached the gargantuan Cedar Rapids ADM plant where many Iowa farmers sell their corn and beans. It reminded us of *Charlie and the Chocolate Factory*. Rising up without windows from the flat plain, it looks like a giant gray alien refrigerator set on end. Enormous pipes and tubes run up its sides like vines. Twenty-four hours a day, seven days per week, long gondola semi-trucks line up like ants at two color-coded gates to unload corn at the black doors, or soybeans at the green doors. The trucks look like miniature toys next to the gigantic walls.

The Cedar Rapids plant is a dry-process mill for corn and soy, one of 149 industrial facilities in the United States operated by ADM, the giant agricultural company worth over $25 billion in 2017. With other plants in Clinton, Iowa; Columbus, Nebraska; and Marshall, Minnesota, ADM has an overall ethanol production capacity in the United States of 1.8 billion gallons per year. Its plant in Velva, North Dakota, can turn out an additional 140 million gallons of biodiesel annually. Ethanol, an alcohol, corrodes pipelines as they are currently built, so it is shipped in tanker trucks to East and West Coast markets. This wastes fuel and adds cost.[5]

As Weber stressed, corn is fuel. Yet profit margins from ethanol peaked in April 2014, dropping from more than two dollars a gallon to just fifty-eight cents a gallon at the end of that year. American farming is still in the energy business despite increased oil and gas from fracking and falling fuel prices, but it may be a transition moment for ethanol. ADM is considering selling its Cedar Rapids plant. Other distilleries have adopted a wait-and-see attitude since government subsidies may hold.[6]

We circled the 300-acre facility several times slowly. The cousin of a small farmer we had met drove into the plant several times a week to unload his giant gondola of cobs and kernels. We considered doing a "Michael Moore"—pulling on boots and Carhartts and riding shotgun with him in order to get inside to see and smell the process up close. We elected instead to email ADM Publicity from a local Starbucks, mention our book, and ask to visit the plant to learn more about their operations. They promptly replied that they do not give tours or allow visitors in their processing facilities, and they could not make an exception for us. "Oompa, loompa!" laughed Jeremy. Perhaps corn-on-corn as far as the eye could see was getting to us.

Other companies were more friendly and forthcoming. A quick 105-mile drive north by northwest from Cedar Rapids brought us to the POET biorefining plant in Glenville, Minnesota.

POET-Glenville takes in 15 million bushels of corn a year to produce 42 million gallons of ethanol. "Yep, that's 1,700 pounds of corn an hour, or three gallons a bushel, 90 gallons an hour, 24/7," Kevin Hobbie, the new plant manager tells us.[7] Hobbie is a solid, folksy man, over six feet tall, and is wearing jeans, glasses, and a faded blue work shirt. He's been operations manager for only four months ("Loving it!") and we are his first "tour."

POET sells to the big gassers and feed companies, but it also supplies Qwik Trips gas stations with E85 fuel throughout the Midwest. The company runs twelve ethanol, or biofuel, plants and strives at each plant to limit water use and reduce Greenhouse Gas Intensity ("GHG" on the company website), using what POET calls "ingreen-uity."

"I've fixed everything in here," says Hobbie, who came up from maintenance. "If it's broke, I can fix it. They figured a mechanic could run the whole plant. I hope they're right!" he laughs. "I cut my teeth in a [meat] packing plant forty-five years ago but packing houses burn you out, fourteen hours a day, seven days a week. That's too much! And when the boss says jump, you ask, how high? But here they ask, what do you need and how can we help you make it better? I've been involved with ethanol for a long time. I like what it stands for. We need to have an alternative to oil. Is this the silver bullet to take us out of the oil age? No. But it's a piece of the puzzle."

Hobbie doesn't break stride as he sweeps an arm across the tall, open room filled with boilers, fermenters, pipes, and tanks. "You can make ethanol out of yeast, out of waste, out of algae. We make it out of yeast and corn. It ferments for sixty-one hours, we drain it, and the liquid goes to what we call the beer well, which is twice the size of the fermentation tank, see there? Oh, don't mind the green soup on the floor"—a goopy green puddle is forming at an outlet valve—"CIP is coming down, that's Clean In Place; they'll do a distillate rinse, a caustic, clean it out. So the stuff in the beer well distills and evaporates and falls through the top with steam coming up from the bottom at around 170 degrees. The alcohol goes off this direction and the bottom goes off that direction, and now we're up to 190 proof, and we need to get the last 10 percent, so we run it through a molecular sieve, and it comes out 200-proof corn squeezins!"

"Impressive!" we laugh at the rapid-fire rendition of how it all works.

"Yep."

"Can you drink it?"

"You do not want to try!"

More laughter. The place smells like a brewery, which in effect it is.

"You know, I lived in the West Indies," says Jeremy, "and folks drank something called over-proof rum, clear as water, and it is evil, evil stuff!"

Hobbie replies, "You put a drop of our squeezins on your tongue and it'll suck the moisture dry!" He points to the little bit of foamy blob on the cement, like an overflow of soap from a washer. "I can tell that's coming from Ferm 8. The bugs are in there and they do their thing, and then we have the syrup, which you can dry and feed to cattle. The co-product is called Dakota Gold, distiller's grains, and it's worth some $150/ton. You see, the solids have the water spun out in a centrifuge, and it's like a wet cow pie splatting. We take out the alcohol and put it back in the dryer."

"How profitable is POET?" we ask.

"What is your idea of profitable? When we are paying over seven dollars a bushel for corn, that's ridiculous! Then we have to add off-spec gasoline to our alcohol, and it costs more than our ethanol. We are at the mercy of the price of corn and the price of gas, both."

"So you are not to blame in the food-to-fuel debate, but a victim yourself?" Jeremy asks.

"You mean, Food vs. Fuel? When was the last time you went out into a field and ate the corn, Professor Jackson? You can't! That's not sweet corn. You have thousands of acres of vegetables north of here. It's the people bashing Food vs. Fuel that have no understanding that the corn in the field is fuel corn. The price of groceries went up 'cause the price of corn went up and that is because the price of gas and oil went up. That's what determines everything."

"Will more ethanol be produced?"

"Oh, yeah! But when the price of corn is so high and the price of ethanol is so low, it's hard to make a dime. This is not the windfall it was ten to fifteen years ago. Now our margins are so freaking close. Makes my job harder."

Hobbie smiles and talks about his hobbies. He has a biodiesel fermenter in his house, and mashes up cattail tubers that he harvests from the pond across the road. Like John Weber, he loves to drive big monster combines with twenty-four-row heads—"rear wheel assist! Two tenders! It's awesome!" He flies in air shows, and fuels his trick airplanes with ethanol. "Me and my partner fly so close together I could walk out on my wing and hop on his. If the product didn't work, I wouldn't risk that." And he's a team roper. "I rope calves cause it's fun. I was on a trail-ride for two weeks, six hundred miles. I was never so glad to get off a horse! A CEO calls me up, wants me to move to London. I say if there's a job in ethanol, I'm ready to go!"

It's long past 5:00 p.m. and we are the only people still in the building.

Hobbie pauses thoughtfully. "Subsidies are on their way out. Farmers are making too much money. Price of corn went up, but so too have inputs. If you are not over 2,000 acres you are a hobby farmer. Farmers are the lifeline around here. We're dried up without the farmers. We are nothing without them. Thanks for coming."

At that time, the use of corn for ethanol was going through the roof. Average annual production of corn grown for ethanol had grown 24 percent a year between 2000 and 2010 and was still going up fast.[8] Sky-high prices threatened food production, because you could make a lot more money growing corn for ethanol instead of for food, with the federal government's blessing. But now the government has to deal with the environmental damage that has been caused by this imbalance.

Theoretically, if managed appropriately, bioenergy could provide a clean, low-carbon alternative to fossil fuels without jeopardizing food production in a world rocketing toward nine billion people. Basic conservation and switching from petroleum to bioenergy could help reduce global warming. But whether biofuels are good for America—or not—is more complicated than our visit to the POET plant suggests. That's because the sustainability of biofuel production depends on a full accounting of the inputs and outputs.[9] It's like bookkeeping. Inputs include the raw materials and energy consumed to make the corn ethanol or soy biodiesel. Outputs include the biofuel product, waste materials, and resulting environmental impacts—principally, carbon dioxide and other greenhouse gas emissions, polluted water supplies, and declining human health. This full accounting is called life-cycle analysis, and by these criteria, ethanol is a terrible choice. It costs more, performs worse, and its production causes more damage to the environment than the petroleum that it suppos-edly replaces, even though reducing greenhouse gas emissions from ethanol production has been scientifically feasible for over a decade.

Biofuel sustainability depends on the source of the "feedstock," or the plant material used for biofuel production; the habitat where the plants are grown; the means of biofuel production; the consequences for human health; and the environmental costs in terms of water con-tamination, toxicity, loss of topsoil, and threats to biodiversity.

The two most important factors in choice of feedstock are whether the plants are annuals or perennials, and how diverse the crop spe-cies are. Root biomass is much greater for perennial species in prai-rie grasslands, such as switchgrass, than for annual crops like corn and soy. This directly increases carbon sequestration in soils, and re-duces how much soil is lost to erosion. Perennials can be harvested while leaving living root systems intact.[10]

Crop diversity is even more important, as David Tilman and his colleagues at the University of Minnesota demonstrated. Using large-scale field and laboratory experiments, Tilman showed that the biological productivity of plant ecosystems increases dramati-cally with the numbers of species in the assemblage. Later, he ap-plied his insights to improving agricultural efficiency and ecological sustainability. Focusing on biofuel production, Tilman laid out plots of degraded and abandoned land and planted them with differ-ent numbers of native grassland species. The results were dramatic: plots with two, four, eight, and sixteen species yielded between 84 to 238 percent more bioenergy than plots with a single species. Al-though all the above-ground biomass was harvested, the annual carbon storage in the plots' soil also increased with species diver-sity.[11]

The overall efficiency of biofuel production can vary widely. Under ideal conditions, the energy obtained from a biofuel (the output) should be a lot more than the energy required to produce the biofuel (inputs from fossil fuels, or ideally from renewable energy derived from solar or wind). Dividing outputs by inputs gives a measure of the efficiency of biofuel production, the "net energy balance ratio," where very large is very good.[12] This ratio is just 20 to 25 percent for corn ethanol and about 90 percent for soybean biodiesel. These percentages might sound pretty good, but they pale in comparison to the efficiency of biofuel production from the sixteen-species degraded soil plots in Tilman's experiments, which reached more than 500 percent. Moreover, when biofuel from Tilman's most diverse plots replaced fossil fuels, reductions in greenhouse gas emissions were six to sixteen times *more* than for ethanol or biodiesel—all without costly and harmful fertilizers and pesticides.

The second factor is the habitat farmed for feedstock production. Tilman conducted his experiments on abandoned farmland. In contrast, industrial-scale biofuel production often takes land out of food production, or clears native forests or grasslands to plant biofuel crops. Conversion from food to fuel raises deep ethical questions in an overpopulated and underfed world. Contrary to Kevin Hobbie's assertions, nearly half of Iowa corn production today (and more before the onset of biofuels) goes to feed the hogs, cattle, and chickens that feed people. Converting American land to produce fuel instead of food magnifies problems overseas because nations that depend on U.S. corn and soybeans for food must clear more land or take other drastic measures to make up the deficit.[13]

Clearing new land for agriculture boosts greenhouse gas emissions far beyond the benefits resulting from the biofuels produced.[14] Compared to gasoline, the net *increase* in greenhouse gas emissions from ethanol produced on newly cleared land ranges from 50 to 100 percent. It takes decades—and in the case of corn, nearly a hundred years—for greenhouse gas savings from biofuels to pay off the initial debt from clearing new land. This kind of ethanol production makes global warming worse. Alternatively, no greenhouse gas debt accrues for biofuels made from perennial prairie grasses grown on previously degraded land. Much more biofuel is produced than from monocultures of corn and soy, and the soil improves to boot.

The third factor for evaluating biofuels is overall production efficiency. This comprehensive figure depends not only on the quality of the feedstock, but also on the sources and quality of the energy used to manufacture the biofuel and, as we learned at the POET plant, the amount of energy used for fermentation and refining. Biofuel production technologies are rapidly improving, but the raw materials used to manufacture biofuels are becoming more diverse and dirtier. Greenhouse gas emissions from new dirty fuels, like the thick, gummy Alberta tar sands or even coal, can be up to ten times higher than for the finest Texas light oil, natural gas, or biofuel from residual leaves and stalks of corn.[15] All of which means that biofuels manufactured with dirty fossil fuels can increase global warming rather than reduce it.

The fourth factor concerns the health costs of fine particulates emitted by different kinds of fossil and biofuels, which have been linked to early death from deep vein thrombosis, and many other health effects. Health costs associated with burning energy-equivalent volumes of gasoline and ethanol made from different sources vary enormously.[16] Surprisingly, the health costs from particulate emissions produced by burning corn ethanol are substantially *greater* than those for burning gasoline, regardless of the energy source used in ethanol manufacture. In contrast, the health costs of burning ethanol made from prairie grasses or corn cellulose debris are considerably *less* than those for burning gasoline.

All of these factors make it clear that biofuel production from corn and soy is little better for the environment than high-quality fossil fuels. Moreover, these biofuels provide only a small fraction of the benefits that could be obtained from more efficient, productive, and environmentally friendly plant biomass. These better sources of biomass are abundant and diverse, and include perennials from degraded lands, reconstructed prairie grasslands, surplus residues from crops and forestry, biomass grown in off seasons between food crops, and municipal and industrial wastes. The biomass potential of all these sources combined is significantly greater than could be produced from corn and soybeans—without requiring expensive fertilizers and pesticides, which themselves are made with fossil fuels—and with the added benefits of minimizing impacts on the environment, biodiversity, food supplies, and human health.

Given all of this evidence, why are we still producing biofuels from corn and soy and making consumers use them in their cars? Because that's where the money is. It's in industrial corn and soy, in the chemical fertilizers and pesticides for corn and soy, and in the massive federal subsidies for ethanol production that prime this pump. A few people have gotten very rich at enormous environmental and human cost. Biofuel production in America has become one of the greatest scams in modern history.[17]

Iowa, as we have seen, is one long, glorious green carpet of soy and corn. But if ethanol is the leading "crop" in Iowa, America's second leading agricultural state, where's the real food?[1]

Table corn comes in at 1 percent of Iowa's total agricultural production. That leaves hogs and chickens. Iowa is the top U.S. producer of both hogs and the hens that lay your eggs. (Nearby Arkansas produces most of the chickens we eat.) Pork is real food: nearly one-third of the nation's hogs are raised in Iowa with exports of $1.1 billion sold mostly to Japan, Canada, Mexico, and South Korea.[2] Iowa does pork really well. Its hog farmers are the best in the world—scientific, efficient, caring. Iowa hogs are happy hogs, washed down and dried off, their manure removed and collected, their feed better than the food of many people around the world.

Over time, the top floor of Ann Budd's house in Iowa City had become our northern field station for this project. Nancy to her friends, Budd, now a leading coral paleontologist, had been one of Jeremy's Ph.D. students at Johns Hopkins University. The location of her home was perfect for many reasons, but when we ventured out to find good wine, local produce, and freshly squeezed orange juice, we found little local food in the big Iowa City supermarkets. They stocked mostly "fresh" produce trucked in from California and Mexico. Nancy had a recommendation for us: "Try the Hy-Vee," a small regional supermarket with sweet corn that was out of this world. One day we asked the manager where he got it. He gave us Marvin Hotz's phone number.

If John Weber represents production agriculture in America's Heartland, then Marvin Hotz's small but profitable farm on the outskirts of Iowa City exemplifies the way it used to be on farms all across the Midwest.

Old Iowa was once covered with orchards and vegetable crops, small herds of dairy cows and cattle, pigs, ducks, and chickens. You can see all of this in paintings by Depression-era artist Grant Wood, in children's books like Harold Felton's *A Horse Named Justin Morgan*, and in songs by farmer-turned-blues-singer Taj Majal. They are invoked by politicians and news anchors—and even an iconic voice of this earlier generation, Rachel Carson. Her epic book *Silent Spring*, first published in 1962, opens with a fable of the idyllic family farm before DDT: "There was once a town in the heart of America where all life seemed to live in harmony with its surroundings."[3]

Besides sweet corn, Hotz supplies Hy-Vee with eggs from free-range chickens. His wife tends the vegetables, and his son, who was at a board meeting for the county fair on the day of our visit, also pitches in. Hotz, more than spry at seventy-two, wearing a red T-shirt and a maize-colored ballcap, soon took us on a jolly ride atop his four-wheeler to see his legendary (to Iowa City customers) fields of "Gourmet Sweet," also known as Hybrid "277A." His main concern as a farmer, he told us, was what made "good food. That's the point, isn't it?"[4]

But first Marvin wanted to show us the chicken house. "We'll just walk over to that chicken house there." Opening the door to the small barn, raucous noises spilled out. Steve said, "Crazy loud in here! "Yeah, it is," Marvin shouted, at the same time gently remonstrating a wayward bird, "Come on, come on, out of the way, come on, out of the way, out of the way! OK, here, here's an egg right here."

"By golly! Look at that!"

The chicken house is a small, clean wooden barn, about fifty by thirty feet, where two hundred Rhode Island Reds feed, hop, and wander about more or less freely. Marvin declares proudly, "They're happy in here."

"This is very different." Jeremy is cryptic.

Steve asks, "That's what free range means? They get to walk around in here as opposed to the little cages?"

"This is free range, yeah. They can run the length of the whole thing."

"I'll be darned . . . Hello guys!" Steve addresses the chickens.

"The thing is . . . a lot of people think that free range should mean just let 'em run all over the farm," Marvin explains, "But, where'd you find all the eggs at then? You know some of those eggs could be three or four days old before you could find the nest."

Steve laughs, "You're absolutely right, I never thought about that."

Now Jeremy speaks up from his farming past. "Actually, where I worked as a kid in Maine, though, the chickens would come in every night."

"Yeah, sometimes they will. The only thing is, if they don't, around here a coon or a fox will get 'em."

"Oh, well absolutely, but in Maine, too, man."

"Good point, Jeremy." Steve steps aside. "Excuse me, chicken."

"We even had trouble with a fox," Marvin relates. "A little boy was out here in the field that one year, and he said, 'Hey, there's red feathers out here in the field, there's red feathers in the field.' So I started watching and once I heard some squawking in the afternoon I got out with the shotgun out here, and the fox was going over the fence and I got him. He had a chicken in the mouth."

Appreciating the story, Steve asks, "And you got him?"

"Got him!"

In his fields of Gourmet Sweet, Hotz plucks a couple of ears and takes out a sharp knife. He slices off kernels and offers a taste.

"Here, try it raw."

The good corn is sweet and tastes fresh. He twists off another ear. The kernels from this ear are doughy. With corn used for animal feed or ethanol, perfection of taste is not the necessary goal. That corn can be more easily worked with machinery, and the bigger the ears the better. On his farm, Hotz explains, they pick by hand using local youth labor, basketball players if possible, for the reach, leaving ears that are still hard or have gone starchy on the ground, " 'Cause I don't want to take any corn into Hy-Vee that's not prime."

Hotz sprays his crops. "They want everything organic, the co-op, and most people today, but I says, you can't have this corn without. You find a worm in it and you'd throw it out. You can't have it both ways. It's either got to be sprayed or it's going to be wormy." He shows us a couple ears that have been sprayed, yet blackbirds have pecked the ends and, "a goddamn bug gets in, because it's opened up."

Pointing out a big line of tiles coming from the fence line, a couple of PVC "tile" strings on the other side of the old slough, Hotz says, "You've got to tile it. If you get a wet year, the tile at least drains the ground out. Otherwise it's mud holes and you get stuck every time you turn around." As with Weber's farm, the tiles drain to nameless ditches, into local creeks and down to the Mississippi. There may be more filtering on Hotz's farm, but it's not intentional. He doesn't apply for conservation easements; he just hasn't gotten around to planting out to the ditches.

Marvin Hotz's bushels are hardly a peck in the mere 3,400 acres of sweet corn planted in Iowa, less acreage than the farmland owned and leased by John Weber.[5] The money and the markets are in fuel, meat, eggs, and processed stuff with nary a thought for vegetables.

Since Hotz was not fully organic and we wanted to visit an organic farm, we asked him whom he recommended. He thought about it, came up with the Schlichtings, Kandy and Jay, up Dysart way. When we pulled into their long driveway, Jay was balancing on a windmill, fixing its jammed gears. Kandy came out to meet us and we talked awhile in the yard before going inside.

"My father said the sweetest perfume in the world is a freshly born pig, little piglets, but I can't even eat a pork chop," she smiles. "I was raised around hogs but I couldn't stand the smell of them. That smell comes back to me." She shrugs as if to say, "go figure."[6]

Kandy and Jay raise organic produce—fruits and vegetables, eggplant and broccoli—as well as the usual agronomic crops of corn, soybeans, and alfalfa. Their farm is in Benton County, just one county to the east of John Weber's farm in Dysart. Counties are important in Iowa. Benton County was named for Senator Benton of Missouri—not for his first cousin Thomas Hart Benton, who painted vibrantly colored Heartland murals like "Hailstorm" or "Ballad of the Jealous Lover," which make the flat land sensuous.

Kandy and Jay's neat brown house, a modern ranch, sits on a groomed lawn under tall poplars behind an extensive "truck garden" of organic vegetables grown for market, all covered with black plastic to control weeds, the good plants poking through. When she was twelve, Kandy used to ride around on horseback de-tasseling seed corn, plucking out "the rogues." If pollen from the wrong varietals blew over and pollenated their fields some corn would get too high, or "just messed up, I guess."

If Harry Truman hadn't beaten Thomas Dewey, Kandy's grandfather, Allan B. Kline, already president of the American Farm Bureau Federation, might well have become Secretary of Agriculture in 1948. Allan Kline was an early monoculturist. He saw the future of corn, believed all the way in corn on corn. When he came back from World War II, "They were thinking they would produce food for the world, you know, and they were very excited about the chemicals," said Jay. But more than half a century later, Jay and Kandy have rejected chemicals to farm organically.

We talk over the kitchen table, the place to talk in Iowa. Jay politely interrupts. He is in his early sixties—like his wife, tall, thin, and fit. "What chemical agriculture will allow you to do is to stop the weeds or other pests and supply nutrients in a hurry so you can get a big crop response. They didn't think about the consequences of what this would do to the soil biology. The nutritional value, and that's kind of a debate, but I think that, you know, the tendency is as corn yields have gone up, the percentage of protein has gone down."

The discussion moves to tiling and on to Roundup, the miracle Monsanto pesticide that kills weeds from the ground up. "What I believe it does," says Jay, "is shut down the weed's defense systems so that soil microbes move up the roots and kill the plant." Roundup is mostly glyphosate (N-(phosphonomethyl)glycine), the herbicide that replaced atrazine when atrazine was banned in Europe in 2004. "Roundup was a silver bullet and it worked for a long time, but in wide application, resistance developed, what you might call unintended consequences," he added.

One unintended consequence, for Kandy, may have been her father's cancer. When he fell ill, neighbors came from all over the county and helped with the work.

Jay and Kandy met in the nondenominational campus ministry in Norman, Oklahoma. "Minister just means servant," explains Jay. "We both came to know Jesus Christ personally when we were in college," Kandy continued. "My father was very excited. We were going to be missionaries so we went to Wheaton College. We ministered in Indiana, Wisconsin, Illinois, and then we decided, because of different things, that Jay should go back and get his agronomy degree." Jay worked for several companies, but his employers all figured that Jay wouldn't be permanent since his father-in-law was Fred Kline of Benton County, Iowa.

As the sun lowered, Jay went back to fixing his mechanical conveyor. He wanted to finish before dark. Kandy took us to the garden and loaded us up with eggplant. She is tall, with swept-back curly blonde hair. As a compliment, Steve asked, "Were you a model?"

"Heavens no!" she looked at him as if he were crazy, then laughs, "Me, no, I'm too shy."

She turned to Jeremy and asked, "After you leave us, how should I pray for you?"

Jeremy paused—no one had ever asked him that before, and he wanted to give a serious answer.

"Pray that we write a truthful book," he smiled.

"OK, that's lovely," said Kandy.

Iowa. Real food is at the margins, where you'll find Christian organic farmers such as Kandy and Jay and dedicated traditional farmers like Marvin Hotz—retro, niche, even spiritual—actually producing very tasty and healthy food. Are they a harbinger of the past, or a portent of American agriculture in an era of climate change?

On the way back to Nancy Budd's house, we detoured a little east and north of Iowa City to the locks at Dubuque.

Eagle Point Park in Dubuque sits 1,100 feet above Mississippi River Lock and Dam No. 11.[7] From Dubuque to New Orleans, the Big Muddy these days does not so much roll on, roll on as glide between channelized banks within the elaborate lock system. Things can be different in spring, when the unruly river may ignore its banks. But most of the year, for 1,010 miles it is pretty much walled in a ditch, with its wilder past relegated to folk songs on Minnesota's *Prairie Home Companion* radio show or in blues numbers sung closer to the Delta. Life on the Mississippi has been domesticated.

There have been levees on the river since the early 1700s, when the French realized the strategic value of New Orleans. Later, Louisiana and Mississippi planters valued the river as a natural highway for transporting cotton.[8] But it was not until the Great Depression that the nine-foot Channel Project was proposed to improve navigation and tame the upper Mississippi from Minneapolis, Minnesota, to Guttenberg, Iowa, and some of its tributaries. Enormous upstream dams built on the Missouri, the Black River, and other tributaries held back the silt that had built the Mississippi Delta since the last Ice Age ended, nourishing swamp and marsh. Fort Peck Lake in Montana, for instance, is the product of a dam on the great Missouri River, and covers an enormous area, 382 square miles. Lake Sakakawea in North Dakota, also a result of a Missouri River dam, covers 479 square miles and is 178 miles long. Sediment retained by the dams would change the volume of the reservoirs over time. Engineers accounted for this, but not for the "sediment starvation" this would cause a thousand miles away, on the Mississippi River Delta.[9]

At first, President Franklin Delano Roosevelt turned down the idea of major impoundments. Conservationists and outdoorsmen, particularly the influential Izaak Walton League, were against it. So were the powerful railroad barons who envisioned their freight monopoly slipping away due to barge traffic. But strong economic arguments were made in favor of the Channel Project. For example, shipping John Deere farm equipment from factories in Kansas to San Francisco would be cheaper down the river and through the Panama Canal than by railroad across the country.

As always, the Army Corps of Engineers was all for building, locking, and damming. Soon they won. Roosevelt approved the scheme in 1933, with the advent of the New Deal. The idea of taming the Mississippi as a mighty make-work project that would employ tens of thousands was very attractive amid the economic crisis. Farmers and anglers, too, would come to depend on the dammed water. Eventually twenty-nine dams and locks on the upper Mississippi would lift barges up and down the four-hundred-foot rise between St. Louis and Minneapolis, from the water a distance of 670 miles. The vast impoundments in North Dakota and Montana held back much of the nourishing silt. Hardening its banks made the river run faster and farther, so that it dropped much of its little remaining sediment into the deep waters of the Gulf rather than in the Delta.

Watching barges pass slowly (very, very slowly) down the river through Lock No. 11 is more exciting than watching corn grow, Steve thought, but not by much. Suddenly something animated Jeremy's interest.

"That's not easy to do, guide fifteen barges side-by-side-by-side! Dude! Put down your sandwich! Those are 5,000-horsepower diesel towboats. They don't use tugs like they do in Panama. Look how the whole tray is then lowered." Jeremy, whose dad had been a master mariner, had made a hobby of studying the movement of huge container ships in the Panama Canal when he had worked at the Smithsonian Tropical Research Institute.

Steve reckoned it was a mighty sight, 120-foot-long towboats thirty feet wide, pushing fifteen barges, each carrying cargo fifteen times greater than a railcar and maybe sixty times that of a semi-truck. Great, if slow, efficiency transpired below our gaze.[10]

"Wow!" smiled Jeremy, shaking his head.

Steve returned to his turkey sandwich.

On our eighty-three-mile drive back on U.S. 151 to Dubuque through Anamosa and Monticello, flat fields of corn descended from hilly land too variegated for big combines and all the toys of industrial production agriculture. The gentle slopes were filled with trees, orchards, and farm animals—Rachel Carson's postcard tableau.

Midwestern water quality is plummeting due to contamination from industrial agriculture. As we saw at Dubuque, dams and levees have transformed the Mississippi and Missouri rivers into a series of tranquil and polluted lakes and wide, placid canals. Most major tributaries have also been dammed and leveed for navigation and flood control. Agricultural runoff, lost topsoil, nutrients and pesticides pouring out of millions of miles of tiles, petrochemical spillage, sewage, and heat from water-cooled power plants and factories all take their toll on the rivers, lakes, and ponds, and eventually the groundwater. Increasingly frequent torrential rains due to climate change (like the storm that hit the Weber farm) hasten erosion, topsoil loss, and add to runoff pollutants. The Midwest is suffering from an acute water crisis.[1] Like the despairing sailor cries out in the *Rime of the Ancient Mariner*: "Water, water, everywhere, and not a drop to drink."

Until recently, the most polluted Midwestern river was the notorious Cuyahoga, which empties into Lake Erie at Cleveland, Ohio. It has caught fire thirteen times since 1858. The largest fire, in 1952, caused millions of dollars in damage. A smaller fire in 1969 ostensibly made the cover of *Time* magazine, but because there weren't any pictures, *Time* used one from the 1952 blaze instead. In the accompanying article, *Time* pronounced the Cuyahoga "dead" and called it a river that "oozes rather than flows."[2]

The *Time* article grabbed public attention and contributed to the huge turnout on the first Earth Day in 1970. It spurred President Richard Nixon to form the Environmental Protection Agency that year, and Congress to enact the Clean Water Act in 1972. Now attacked by people who don't remember what it accomplished, the Clean Water Act worked miracles. The Cuyahoga no longer burns, although there is still a musical Burning River Fest every year. Oysters are also making a comeback in New York Harbor. Like the Clean Air Act of 1963, it is a landmark of twentieth-century conservation.[3]

But the law was no panacea. Cuyahoga is still polluted with high levels of *E. coli*, and pollution throughout the Midwest not only persists, but is rapidly increasing. Laws only regulate point source pollution, that is, pollution from identifiable sources like power plants, chemical factories, and sewage plants. In contrast, nutrients and pesticides from industrial agriculture, which spans tens of thousands of fields across the Heartland, are defined as nonpoint source pollution, and therefore are unregulated. This loophole is a major weakness in the Clean Water Act: it fails to regulate urban runoff, and it permits dangerous poisons to enter drinking water. Old, deteriorating infrastructure and industrial pollution in rivers have compounded these problems and caused the drinking water tragedy in Flint, Michigan.[4]

Getting statistics on the extent of agricultural water pollution is a bit of a nightmare. The U.S. Department of Agriculture (USDA) publishes a detailed compendium every five years, the most recent of which is the *2012 Census of Agriculture*. For each state, crop statistics are given in terms of the number and size of farms, numbers of acres planted with various crops, and numbers of bushels produced. The report excels in measuring production, but is vague about measuring substances with potentially harmful effects—a pattern reminiscent of the department's historical acceptance of DDT and other pesticides banned today.[5]

Monitoring potential environmental hazards is done instead by the U.S. Geological Survey (USGS), in the Department of the Interior. The USGS National Water Quality Assessment Project produces detailed maps, data tables, and reports pertaining to threats to water quality, and a cursory examination gives a rough indication of how much pesticide and fertilizer have been used. Additional information from the Fertilizer Institute states that the average amount of fertilizer used is down to 1.6 pounds of nitrogen, phosphorus, and potassium nutrients per bushel of corn. Soybeans can take up nitrogen from the soil directly and through their symbiosis with rhizobia bacteria in their nodules, but they feed heavily on fertilizers. Farmers use roughly two pounds of combined nutrients per bushel to boost soybean production to a hundred bushels per acre.[6]

Iowa leads in most acres in corn plus soy, most bushels produced, and greatest acreage fertilized and treated with herbicides. The numbers are staggering: nearly 25 million acres are planted in corn and soy, most of it GMO. Commercial fertilizers are spread on 80 percent of these crops, which annually yield about 2.9 billion bushels of corn and soy combined. Based on an average application of 1.5 pounds per bushel, that translates into roughly two million tons of nitrogen- and phosphorus-rich fertilizer injected into Iowa fields every year.[7]

Second in grain production and acreage is Illinois; Minnesota is third; and Nebraska is fourth. Each produces over a billion bushels of corn and soy annually and uses over a million tons of fertilizer. Eleven additional states in the Missouri and Mississippi drainages produce another four billion bushels and use three million more tons of fertilizer. That all adds up to about eight million tons of nutrient-rich fertilizer a year.[8] Its use has exploded since 1950, and now there's hardly a drop of water in the Midwest that isn't laced with high concentrations of nitrogen, phosphorus, and potassium. The Great Lakes and the Gulf of Mexico have a bad nutrient problem.

One source of data on herbicide application estimates that on every acre, every year, two cups of liquid Roundup, the weed killer used on genetically modified soy and corn, are applied, along with numerous other herbicides and fungicides. This means that virtually all soy and corn (and farther south, cotton) are treated with glyphosate, the principal herbicide in Roundup. The total cropland in corn and soy for the fifteen states described earlier is 140 million acres: at two cups per acre, this works out to an astounding 17 million gallons of Roundup per year. That's the equivalent of twenty-six Olympic swimming pools filled to the brim with poison for plants. If we use USDA estimates of pounds per square mile for high usage areas, this comes to more than 19 million pounds.[9]

Leaving aside the consequences of all that poison for environmental and human health, this level of chemical intervention is way too much for sustainable agriculture. Herbicide-resistant weeds are invading fields in abundance, and wreaking havoc across the Corn Belt. To make matters worse, new "superweeds" such as horseweed, giant ragweed, and pigweed are rapidly evolving resistance to Roundup and maybe even like the stuff. Pigweed grows an astonishing three inches a day to a final height of more than seven feet, so it can easily crowd out corn and soy.[10]

Superweeds escalate the agricultural arms race. Nearly half of U.S. farmers surveyed reported glyphosate-resistant weeds on their farms in 2012, with a 51 percent increase between 2011 and 2012. The diversity of superweeds is also increasing. In 2010 the *New York Times* mapped the numbers of glyphosate-resistant superweeds in each state. The epicenters are Arkansas and Kansas, with six different resistant weeds, followed by Iowa and Ohio with five. Chemical industries like Monsanto are countering this by reintroducing synthetic auxin herbicides marketed as dicamba, which had been shelved because they damage corn and soy, and developing GMO crop varieties that are resistant to both glyphosate and synthetic auxins.[11]

This expensive, short-term fix will inevitably breed the next generation of superweeds. It will also pose a grave threat to organic or other nonresistant crops being raised by nearby farmers who for one reason or another have not adopted auxin-resistant strains. This situation raises profound ethical questions of environmental justice: why should some farmers have the right to plant crops that will financially damage their neighbors? As we go to press the issue is becoming explosive because neighboring farms are suffering adversely in Arkansas and other areas of the Corn Belt can't be far behind. Tempers are running high and a dispute between two farmers about the herbicide resulted in a shooting death and charge of murder. Arkansas has put a moratorium on dicamba for 120 days.[12]

These developments were predictable, indeed inevitable. Rachel Carson chronicled DDT-resistant pests in *Silent Spring* half a century ago.[13] Nevertheless, pressures to use chemicals without adequate testing, or without otherwise taking into account adverse effects, will be enormous. As more and newer chemicals are dumped onto fields in the future, the prospects for clean, healthy water decrease more and more.

Toledo, Ohio, lies ninety-nine miles west of the Cuyahoga River on the southwestern banks of Lake Erie. Its freshwater supply was completely shut off on the first weekend in August 2014 to protect public health. Nearly half a million people suddenly had to depend on bottled drinking water with no water to cook with or bathe in. The ban was lifted after two tense days, but warnings are still issued from time to time because the problem won't go away anytime soon.[14]

The poison that shut down Toledo is the biotoxin microcystin, which is produced in abundance by a microbial cyanobacterium *Microcystis*, and a blue-green alga, *Anabaena*, which thrive in the nutrient soup of dissolved fertilizer from Midwestern farmlands. These and other microscopic organisms that produce toxins comprise a major threat to all freshwater systems and the people who depend on them for water. Microcystin causes headaches, diarrhea, and vomiting, and accumulates in the liver, where it can do major damage, sometimes causing death. It's often fatal to dogs, livestock, and other animals. Boiling water poisoned with microcystin doesn't reduce its toxicity. Remarkably, there are no federal standards for safe levels of microcystin or guidelines for its measurement. States have been left to fend for themselves.[15]

Toledo was unlucky. A relatively small cyanobacterial bloom formed right over the city's freshwater intake from Lake Erie. But the largest bloom ever recorded in Lake Erie had occurred just three years earlier, after record-setting spring rains. Anna Michalak and twenty-eight colleagues described what transpired.[16]

The 2011 bloom began to form in mid-July. Initially it covered about 230 square miles. Then it spread a hundred miles eastward, until, by October, it covered almost two thousand square miles—a toxic area larger than the state of Rhode Island. Surface-water toxins reached almost twenty micrograms per liter, half the World

Health Organization danger level for human health, and could well have exceeded this limit in places.[17] Satellite imagery showed that the bloom was more than twice the size of the 2008 bloom, then the largest on record, and seven times greater than average blooms over the previous nine years. Like "storms of the century" that now happen every five to ten years, widespread toxic blooms are the new normal for Lake Erie and other lakes throughout the Heartland.

Why are these events occurring? Cyanobacteria were always there, but they were held in check by lack of nutrients, and by freshwater clams and other suspension-feeding animals that filtered them from the water for food. But overdependence on fertilizers, widespread use of drainage tiles, and wholesale destruction of wetlands that used to capture runoff destroyed that balance. Enormous quantities of nitrogen and phosphorus now flow freely into watersheds and fuel cyanobacterial growth. High concentrations of dissolved phosphorus (DSP) are particularly bad because phosphorus stimulates the growth of cyanobacteria in lakes.[18]

Michalak and her colleagues identified three culprits in the unprecedented 2011 bloom.[19] The first was harmful farming practices. Large amounts of fertilizer were being applied on the surface of fields in autumn instead of being injected into the soil. Conservation tillage, where crop residue is left on the fields after harvest to reduce soil erosion and runoff, reduces soil loss and is generally good practice—but not when the residue is laced with excess nutrients. The 2010 crop had been good, and farmers had fertilized heavily.

The second and third culprits relate to climate change. Patterns of frequent torrential rains interspersed with episodes of drought had increased by a third in the Midwestern Corn Belt between 1958 and 2007. Gullies had formed and expanded, signaling increased soil erosion and runoff, which carried excess nutrients. In 45 percent of Iowa townships, total soil erosion averaged twenty tons per acre between 2002 and 2010, but in a few places it was as high as a hundred tons per acre. Spring rainfall in 2011 broke all records, causing as much as a billion dollars in agricultural losses and inflicting lasting ecological damage from washed-out soil and nutrients. Increasing temperatures favor cyanobacterial growth. Warmer weather also decreases vertical mixing in lake waters and allows algae to accumulate on the surface, and in 2011, this problem was exacerbated by exceptionally still water.[20] In a nutshell: the bloom was caused by too much fertilizer, excessive rain, soaring temperatures, and lake waters that were too calm. Climate change is accelerating, so most of the variables in that list will continue to cause a problem. Unless farmers cut fertilizer use to a minimum, toxic blooms will get bigger and worse.

Enter an invasive species. Zebra mussels (*Dreissena polymorpha*) now carpet much of the Great Lakes, and their presence benefits cyanobacteria.[21] In sixty-one Michigan lakes, with and without the mussels, lakes with mussels had higher concentrations of *Microcystis* and its toxin, microcystin. Zebra mussels filter huge quantities of lake water for plankton, but they apparently don't like *Microcystis* and consume other algae instead. Zebra mussels' food preferences allow *Microcystis* to proliferate as other algae are reduced. In this way, the tiny mussels become another factor responsible for the spread of cyanobacteria and their poisons.

Although the Great Lakes get most of the attention, cyanobacteria populations have exploded throughout the Corn Belt. Of lakes monitored in Illinois, Iowa, Minnesota, and Wisconsin, 40 to 59 percent tested positive for microcystin, often at dangerous concentrations. The World Health Organization publishes the risk associated with recreational exposure to cyanotoxins: less than 20,000 cyanobacterial cells per milliliter is low risk; 20,000 to 100,000 cells per milliliter is moderate risk; and more than 100,000 cells per milliliter is high risk. In Iowa in 2010, all but one of 26 lakes that provide drinking water had *average* summer cell counts reflecting moderate to high risk. But averages obscure exceptionally high levels. In the same study, 29 of 32 lakes sampled over a longer period, 2006 to 2010, had cell counts in a single test well above the danger zone—up to a whopping several million cells per milliliter. Although Illinois, Minnesota, and Wisconsin lakes were considerably healthier (38 to 74 percent of lakes tested were low risk), a significant number ranked from moderate to high risk.[22]

These lakes provide almost all the drinking water in the Midwest, so this contamination poses an enormous—and increasing—health hazard. The 2012 Environmental Working Group report used data collected more than a decade earlier by the USGS that also clearly highlighted the risks to public health; other systematic analyses highlighted the problem as well. Nevertheless, by the time of the 2012 report there were still no legally established limits for the amount of cyanotoxins in drinking water. Consequently, the Environmental Protection Agency (EPA) can publish guidelines for safe limits of cyanotoxins, but is powerless to shut down poisoned drinking-water systems that are a hazard to public health.[23]

Excess nitrogen itself poses another serious health threat to drinking water. Well water and water systems in small towns throughout the Corn Belt commonly have nitrate levels well above the EPA danger limit. Nitrate in water is converted to nitrite in the human body, where it interferes with the blood's ability to carry oxygen to body tissues. Blue baby syndrome in infants can be fatal, and it is clearly related to nitrate in drinking water. High nitrate levels may disrupt thyroid function in a fetus and double the risk of thyroid cancer in elderly men and women. Breakdown byproducts of disinfectants used to purify polluted drinking water have also been associated with an increased risk of bladder cancer and other ailments.[24]

Finally, let's look at the herbicides that the EPA has long insisted are safe. Glyphosate, the major ingredient in Roundup, is considered safer than atrazine, the previous herbicide of choice that was banned in Europe in 2003 but is still legal in the United States. Yet concerns about glyphosate have been growing, to the extent that in 2015 the World Health Organization announced that it is "probably carcinogenic." The organization reversed its advisory one year later, but there are questions about scientific objectivity. Nevertheless, the most recent and exhaustive summary of the issue posits significant adverse effects on human health. Much of the relevant data linking glyphosate to cancer has been around for at least twenty years.[25]

It is also increasingly apparent that combinations of chemicals can be more dangerous than single ingredients by themselves. This amplification effect has been shown in numerous studies of commonly used chemical mixtures, even though testing each chemical only in isolation is standard protocol. In several cases, a mixture's toxicity was twenty times greater than that of individual components in isolation.[26] It is worrisome that inert ingredients may be safe by themselves, but magnify dangers from active ingredients.

Growing concerns that glyphosate is "probably carcinogenic" have set off increasing efforts to ban Roundup in Europe and the Americas. Under intense pressure from the manufacturer Monsanto, and unable to reach a unanimous decision, the European Union gave Roundup a last-minute reprieve in 2016. Nevertheless, the Netherlands has banned its noncommercial use and other individual members of the European Union are moving forward with various measures, including listing glyphosate as a carcinogen and requiring labeling to that effect. Major supermarket chains and other businesses in the Netherlands have also stopped selling glyphosate products. As we go to press, the European Union has called for glyphosate to be banned by 2022, but after much debate approved by a qualified majority the licensing of the weedkiller for another five years.[27]

The U.S. Environmental Protection Agency has expressed concerns about Roundup but has not restricted its use. The California EPA, however, has moved to list Roundup as a carcinogen and, as of January 2017, the state cleared a preliminary hurdle to require a cancer warning label on Roundup products. As it did in Europe, Monsanto is fighting back but has lost the first round in court. With 12 percent of the U.S. population and thus a large fraction of Congressional representatives, as well as the sixth largest economy in the world, California has a lot of political clout and often dominates the national environmental agenda. For instance, the automobile industry failed to turn back the Golden State's stringent fuel and emissions standards, which its representatives had challenged in court because manufacturers couldn't afford to produce two kinds of cars: efficient ones for California and gas guzzlers for the rest of the country. That battle continues with California's ever more stringent emissions requirements.[28]

Meanwhile, 250 million pounds of Roundup continue to be dumped on fields across America every year.[29]

It's not just human health that's threatened by "corn on corn" farming. In 1962 *Silent Spring* taught us about the folly of depending entirely on chemical control for crops' success. Vital pollinator species including bees and butterflies are now being threatened by the use of pesticides and the destruction of essential habitats. For example, the sharp reduction in milkweeds and the population collapse of monarch butterflies, which depend on them, has been closely linked to Roundup. Monarch ecology is enormously complex, involving long-distance migrations and threatened overwintering grounds. Nevertheless, the loss of monarchs' basic food supply over vast areas has been a major factor in their demise, as evidenced by the recovery of local populations where milkweeds have been reestablished. Neonicotinoid pesticides also face bans or restrictions because of compelling evidence, presented in 2015 in the journal *Nature*, of their fatal effect on honeybees, which are in precipitous decline across the United States. (The European Union had already banned the use of three neonicotinoid chemicals in 2013.) Honeybees pollinate a vast number of agricultural and wild plants. But in the United States, most seed corn and soy are still impregnated with neonicotinoids before planting, and there are concerns about widespread chemical dispersals via the wind.[30]

Deeper analysis may reveal more problems that no one wants to talk about. Just like the harmful effects of tobacco on human health, of chlorofluorocarbons on the ozone hole, and of acid rain caused by coal-burning Midwestern power plants on Northeastern forests, we continue to use these chemicals until it can be proven beyond a shadow of a doubt that they are harmful—instead of heeding the so-called precautionary principle, which would require manufacturers to prove that chemicals are safe before selling them.[31] In the long run it would be much easier, cheaper, and safer to follow the precautionary principle than to remove dangerous materials already in use and pay for remediation.

We have known for at least twenty years that industrial farming practices harm human health and that attempts to remediate the damages have posed enormous financial burdens on the public. The costs for making drinking water safe in the Corn Belt are already staggering. The Des Moines Water Company sued three rural counties in Iowa for failing to reduce nutrient runoff, but lost the case in the Iowa Supreme Court, which ruled in favor of the county water districts. The city water company's frustration is easy to understand. The costs of cleaning up the drinking water for a few thousand people or fewer range from several hundred thousand to over a million dollars, roughly $100 to $1,000 per person. Reverse osmosis plants, the latest technology, may cost $20 million to service the same number of people, although the costs are coming down. Scale that up to Lake Erie and the rest of the Great Lakes, and tens to hundreds of billions of dollars may be needed to make Midwestern water safe to drink.[32]

That's the real price of corn.

PART II

Coast

Jeremy raised his eyebrows and nodded solemnly to R. Eugene ("Gene") Turner, seated beside him in the loud bush plane. Steve was in back. We all had our "ears" on, those big muffs with jutting stick mikes that you wear in helicopters and small planes. It was impossible to talk.

Turner, professor of oceanography and coastal sciences at Louisiana State University, is one of America's leading experts on marsh and wetlands. His work can be controversial. Marshes—what destroyed them, how to bring them back, whether some can ever be restored—are contentious topics. Turner had driven down to New Orleans from LSU in Baton Rouge to guide us through an aerial view of the Delta.

Seen from the air, the Mississippi Delta looks like a gigantic cancerous lung, excoriated with ten thousand miles of oil and gas canals, widened and widening over decades into a vast disassemblage of open water. A few misplaced piles of leftover earthen refuse from the excavations, called spoil banks, poke through like the backs of dead alligators. It's a sickening sight.

This vast open wound of disembodied grass islands in mostly open water has taken the place of what was once a wide, waving sea of grass that buffered and protected the city of New Orleans from storms. All of a sudden, we realized the scale of utter destruction of the natural environment of southern Louisiana. It's dramatic. Huge gaping watery holes are everywhere.

Before our flight, Turner, a thirty-five-year resident of Baton Rouge, had spread out a bundle of maps on the cement floor in the little lobby on the corporate side of Lakefront Airport at the edge of Lake Pontchartrain. "All this was marsh in the 1930s," he explained, "until it was dug out and channelized by the oil companies. Back then, companies thought marsh was worthless, and they would run boats and equipment over it. But there were also oyster reefs further up the delta, closer to New Orleans. These oyster reefs did have understood value back then and were spared."[1]

"That shows some consideration," Jeremy said.

"Yes, they wanted to do the right thing," replied Turner, "but values were different then, and the value of marsh to preserving land and buffering cities against hurricanes was not understood very well."

We stepped outside, climbed into the single-engine Cessna, and went up with a whoosh on an updraft. Light ricocheted from the few downtown skyscrapers as we banked over the lake. Behind us lay the famous Ninth Ward, where levee walls had broken during Hurricane Katrina, while to our left was the unmistakable top of the Superdome, where so many New Orleans residents had been trapped. Our pilot was skinny, tall, bespectacled, and competent. The day was sunny and clear. Turner pointed down immediately, but with the vibration, the thrum of the propeller—it was just too loud for us to hear even with the fat earphones we had been given. He kept pointing and shouting with a few smiles and much difficulty. Soon the plan became to review what we saw back on the ground.

Here's the hundred-mile loop we flew: first we headed east over the Mississippi River Gulf Outlet Canal (MRGO), or what locals call "Mr. Go"; then we went farther east and southeast across St. Bernard Parish and the Breton Sound Estuary; next we passed over the Alliance Refinery and, farther to the south, over Woodland Plantation (soon to become our raucous southern field headquarters) as well as nearby expanses of grass and water. Turner would soon take us there in his LSU boat to demonstrate the scientific value of jumping on marshes, soggy and hard. The Grand Bayou Indian fishing village lay close by, and past it, toward Bay Baptiste, we could see from the air the islands and coastline lined with useless booms left over from the BP spill. Then we flew north on the west side of the Mississippi River, above Jean Lafitte National Historical Park and Preserve, where Turner would soon show us good uncut, dendritic marsh still standing—in other words, half-floating. A bit west over open water, we passed Delta Farms, flew over the Davis Pond diversion, and eventually came back to the river again at Maurepas, where we spied curious tracks that looked like spokes emanating from a giant crop-circle, or glass-skull alien wheel, later explained. Flying north to Hammond, we flew over a sewage treatment plant that used wetlands with the goal of building marsh. But instead, Turner contended once we landed, this influx of upstream nitrogenized water had weakened the roots of the grasses until, in the latest hurricane, the thinned vegetative cover blew away like a bald man's toupee. Not surprisingly, the controversial sewage treatment plant had acquired derogatory nicknames, Turner told us, such as Lake Poo Poo or Lake Commode. This observation elicited smiles, but the entire tableau was not funny.

Seen from above, it is possible to imagine the river as it was before Europeans arrived.[2] The natural levees often found beside rivers and bayous eventually became the high ground for colonizers. First, plantation roads were dirt, then paved. A lattice of developed roads now spreads out from New Orleans like the fingers on an open hand. Today I-90 cuts a path across the base of the original fingers, and I-10 extends across the palm.

At the north end of our seventy-minute flight, the Mississippi lay below us almost as mythically muddy and wide as it once was. Here we could imagine seeing far upriver from Louisiana, past Iowa, all the way to little Lake Itasca, in Minnesota. But that's sheer fantasy. Today dams strangle tributaries like the Missouri to form reservoirs as big as small states. Twenty-nine locks on the upper Mississippi interminably lift slow-moving strings of massive, coupled barges carrying coal, chemicals, grain, and other heavy cargo. To the south, more than 1,600 miles of levees and floodwalls constrict the main river alone. Once the great Mississippi may have been "Old Man River," but there has been some gender bending since Huck Finn drifted down it: the Old Gal is now well-corseted. True, she cast off her garments and went wild in the Great Flood of 1927. There were later extreme events, too, notably the high-water year of 1973. But the problem isn't just really bad weather. Geology set the groundwork for the river's unruly behavior.[3]

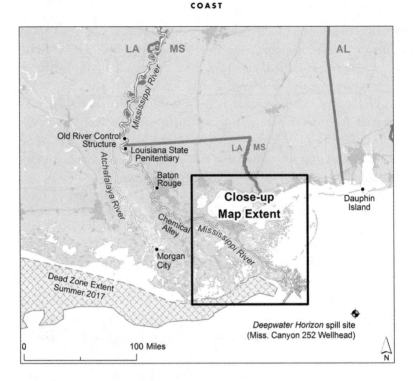

Mississippi Delta and southern Louisiana with the geography of
the Lower Mississippi and Atchafalaya rivers, the Old River Control
Structure, and Chemical Alley.

Two sets of geological factors explain the willfulness of the Mississippi, why it takes so much effort to keep it in place and why in the end it will win.[4] The first set comprises its history, and the composition of land beneath the riverbed and delta. The second set encompasses the way the delta has switched back and forth and up and down. This relates to fluctuations in sea level and the river's compulsion to find the shortest route to the coast.

The Mississippi drains nearly half of the continental United States. Its major tributaries include, to the east, the Ohio River, which drains the Appalachians and sends the most water downriver, and to the

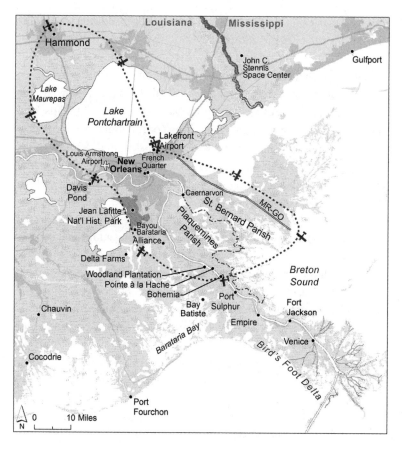

Route of our overflight of the Mississippi River Delta region.

west, the Missouri River, which drains most of the Rockies and once delivered most of the sediment that built the Delta and kept it above sea level. The river's main course dates back more than 70 million years to the Cretaceous period, but it became the principal drainage of the central United States only about 15 million years ago.[5]

The current situation started to evolve about three million years ago, when Northern Hemisphere glaciation started and a massive

North American ice sheet began to form. The ice blocked the north-ward drainage of the Missouri and Ohio rivers into Hudson Bay and turned them southward instead. At this point sediment accumulation downstream began in earnest along a roughly 250-mile stretch of the Louisiana and Mississippi coasts. Today that sediment layer is more than 1,500 feet thick along the latitude of New Orleans, and more than 13,000 feet at the edge of the continental shelf. The sediments are soft, unconsolidated, and easy to shift and deform. They also contain high concentrations of organic matter, which is the source of the rich petroleum deposits that support a quarter of U.S. oil and gas production. But the entire Delta region is inherently un-stable due to the weight of all that sand and mud, which compacts the sediment and pushes the Delta downward, causing additional subsid-ence. This is crucial for the future of the coast.[6]

The sea level has fallen and risen with glacial advances and retreats. During the last interglacial period, 125,000 to 80,000 years ago, the sea level and geomorphology of the Delta were similar to conditions today. But then the ice advanced again. Eventually it covered all of Minnesota, Wisconsin, Iowa, Illinois, New York, and New England; most of the Dakotas, Indiana, and Ohio; and about a quarter of Ne-braska, Missouri, and Pennsylvania. During this era, the sea level dropped about four hundred feet (there were numerous wobbles up and down), and the ocean retreated all the way to the edge of the con-tinental shelf, as the Mississippi carved a deep valley all the way down to sea level.

Then, about 11,000 years ago, all hell broke loose. Temperatures warmed and the glaciers suddenly began to retreat—so rapidly that the sea level returned to present levels in about five thousand years. It must have been a wild time. The whole incised valley filled up again, first with coarse gavel and sand from the initial torrents of gla-cial meltwater cascading down the river, then with finer silts and muds as the front of the Delta retreated to its present position far away from the edge of the continental shelf.[7]

Ever since, the Delta has gyrated back and forth across the Loui-siana and Mississippi coast about every one to two thousand years.

Seven thousand years ago it ran way west of Baton Rouge. Then it moved eastward about fifty miles until, about 3,500 years ago, it long-jumped 150 miles eastward, offshore of St. Bernard Parish. Five hundred years later, it jumped westward about 75 miles until, sliding east again to its present position, it formed the Plaquemines-Balize Delta about 1,300 years ago. All of which means that the Mississippi has switched courses five times in the last seven thousand years, in paths that span more than two hundred miles. You can see these former deltas from the air. It looks like a snake thrashed its tail from side to side.[8]

Now the river is itching to jump again into the Atchafalaya River channel and enter the ocean 150 miles to the west. This courtship has been getting serious over the last five hundred years, and the only things standing in the way are human arrogance and the Army Corps of Engineers.

The Atchafalaya ("Long River" in Choctaw) formed the southernmost portion of the Red River until the 1400s, when the naturally meandering Mississippi River cast an oxbow to the west that intercepted the Red River at Turnbull's Bend near Angola, Louisiana. What remained of the Red River became the Mississippi's tributary. The Atchafalaya, shunted off from points west, was born.[9]

The oxbow eventually proved an irritant to navigation, until Captain Henry Shreve cut a channel connecting the upper and lower Mississippi River more directly. This channel, called Shreve's Cut, substantially changed the flow of water in both the Mississippi and the Atchafalaya. The oxbow transected by the cut became known as the Old River, and the relative heights of the Red River and Mississippi determined where the Old River would send its water. When the Red River was high, and the Mississippi low, the Old River could flow south down the Mississippi; when the Red River was low and the Mississippi high, it could flow west down the Atchafalaya. Over time, the flow balance between the Atchafalaya and the Mississippi reversed, so that starting in 1880, most Old River water went west to the Atchafalaya, whose channel grew deeper and wider. By 1950, the Atchafalaya was receiving 30 percent of the Mississippi's water flow.

Another event like the Great Flood, however, could cause the water to switch paths again, particularly since the western route to the sea is shorter and steeper.[10]

To keep the Mississippi flowing as it is today, the Army Corps of Engineers imposed strong discipline. They shored up the banks with massive amounts of dirt and steel and entirely blocked off the Old River, except for a system of floodgates, locks, and diversions that permitted barge traffic between the two large waterways. The Corps finished construction in 1963 and named the entire system, prosaically, the Old River Control Structure (ORCS). In the mid-2000s, Steve paddled the Mississippi south of Clarksdale and the Atchafalaya for a series of popular science articles. He toured the ORCS and stood on its steel catwalks after another high-water spring, when the entire structure vibrated and rattled like an old freight train. The Atchafalaya still runs faster, deeper, and more directly to the Gulf of Mexico.[11]

During times of heavy flow, the Army Corps worries about scouring, whereby powerful underwater currents could hollow out the riverbed beneath the pilings. If ORCS tears out and backup levees and spillways give way, the Mississippi would permanently join the Atchafalaya and roar down to the Gulf of Mexico at Morgan City, eighty-seven miles west of New Orleans. Personal and business property downstream would be devastated, not to mention the potential loss of life if there wasn't sufficient warning (for an analogy think about the near disaster of the Oroville dam in California during the early 2017 flood rains). At the same time, the flow down the existing Mississippi channel past Baton Rouge and New Orleans would slow to a trickle. This would allow saltwater to intrude way up the Mississippi riverbed past Baton Rouge, shutting down all the plants and municipalities along the way that depend on abundant freshwater intake.[12]

A failure of ORCS would mean a major blow to the U.S. economy. Louisiana has the second largest oil-refining capacity in the United States, with seventeen refineries processing three million barrels of crude oil per day.[13] Crude oil from foreign countries travels up the river in tankers, whereas regional petroleum and natural gas, often from far out in the Gulf of Mexico, travel through a web of pipelines to the refineries and processing plants along the Mississippi. Besides energy production, the petrochemical plants convert hydrocarbon feedstock into a panoply of chemicals. The nitrate fertilizers that are spread over farmland and pollute drinking water to the north are manufactured from natural gas in facilities near New Orleans. All these refineries and chemical plants need water for mixing and cleansing, and they discharge wastewater into the river.

In fact, the 85-mile stretch of the Mississippi between Baton Rouge and New Orleans contains 25 percent of the nation's production capacity in the combined petroleum and chemical industries. This petrochemical corridor has come to be known as "Cancer Alley." Residents and visitors noticed first the dense, putrid air that deposited grime and seemed to cause irritated eyes and sore throats. Then, in 1987, elevated instances of cancer were noted in St. Gabriel, a small, largely African-American community on the Mississippi River twelve miles south of Baton Rouge. But that was not all. Between 1985 and 1988, in the town of 2,100 people, sixty-five women had seventy-five miscarriages. Similar rates were soon discovered in towns all along the corridor. Thirteen years later the Louisiana petrochemical industry was producing close to 9.5 *billion* pounds of toxic waste, more than two thousand pounds of waste for each citizen of the state. Today over 130 industrial facilities, incinerators, and landfills, including seven of the largest polluters in Louisiana, are located on Cancer Alley.[14]

Industry and government, however, were not convinced of the connection. A 2004 study conducted largely by Shell Oil Company scientists using demographic statistics found that, overall, cancer mortality "in the Industrial Corridor does not exceed that for the State of Louisiana, although Louisiana's rates were higher than those of the US, on average." Race was identified as contributing to cancer risk in this study, as were low income, obesity, and smoking, and these factors cast doubt on industry culpability.[15]

Recent research linking the distribution of airborne contaminants, particularly formaldehyde and benzene, to neighborhoods with different demographics found that those in "Cancer Alley, as a whole, have higher exposure [to airborne toxins] and cancer risks." According to these studies, some 46 people per million can be expected to get cancer in Cancer Alley, compared to 30 per million in the entire United States. Income disparity exacerbated risk, which was 12 percent higher in poor communities like St. Gabriel than in wealthier ones. The risk of getting cancer was also 16 percent higher in predominantly black communities. Furthermore, risk compounded as the area income declined and the proportion of African Americans increased. A sardonic joke of the time noted, "People in the Garden District [of New Orleans] get bottled water, while the people in the ghetto just get cancer."

Today, people in St. John the Baptist Parish who live near a facility that manufactures neoprene "have the highest risk of air pollution-caused cancer in the country." The risk in La Place is "800 times higher than the national average."[16]

Sobered by what we had seen from the air, we touched back down at Lakefront Airport, and repaired to a nearby bar. This joint was an appropriate place to discuss the ravages of the marsh. During Katrina it had lost electricity and was flooded, basically ravaged by the storm, which was not surprising since the dive was below street level, with basement windows poking out for daylight. The owner told us that he had camped inside for days, giving free booze to people who wandered in, but also sleeping with a shotgun in case passersby wanted more than a drink.

"Gene, the flight blew me away!" began Jeremy. "It was an epiphany!" and to the waitress, "I'll take a Corona, thanks. Steve?"

"Albita."

"Gene?"

"Iced tea, please."

Jeremy continued, "The overflight was [a look at] the endgame of the destruction of the last one hundred fifty years. I get it now. Thanks for a vision of the apocalypse!" He raised his glass.

"Um, well, thank you." Embarrassed but chuckling, Turner clinked his tea with Jeremy's Corona.

"But Gene, what were those spokes on the ground at the very end of the flight," Steve asked, "before the bridges over the swamp toward Baton Rouge?"

"Those were the tracks of giant cypress trees being dragged across the marsh about a century ago," Turner explained. New Orleans never had protective mangroves like Miami or coastal Thailand. But it once had huge, thousand-year-old cypress, the iconic southern trees often draped with Spanish moss. Those trees were valuable, like the Louisiana egrets slaughtered for women's hats.[17] Like redwood from the West Coast, cypress doesn't rot and is resistant to insects. They were clear-cut to provide lumber for building New Orleans outward, and also for export to the north. Teams of oxen and early steam and gasoline tractors hauled the logs out. Often swamps were drained to make this possible, and levees in places kept the land dry. But the soil oxygenated and compacted, so impressions of the tracks stayed, drawn into the soil, visible from the air like the Nazca lines in Peru, which were identified from small planes.

"Oxygenated?" Steve asked.

"Remember," Turner admonished, "this is not mineral soil. This is organic soil, like mulch, and when it dries out, it compacts and sinks."

He explained that the Delta farm we saw was a failed 1915-era agricultural experiment, which failed again in the 1960s. The farm was built on drained marsh so that as the organic soil became oxygenated it would compact down below sea level. This was fine until the levee surrounding the farm collapsed during a storm and water entered to flood the drained and compacted marsh. This flooding was the last. It's now open water.

Turner pulled out his laptop and drilled down through a time-scaled overlay he had created to show us how the old drainage ditches showed up on the farm. He used the same remarkable satellite tools that archeologists have employed to trace the foundations of Roman villas.

"I got a little confused with MRGO versus the Caernarvon Diversion . . . ," Steve began.

"They are completely different!" said Jeremy, "and in completely different places!"

"What I mean is, not the long, open part of MRGO out in the Delta, but the part in New Orleans."

Turner first made sure Steve understood that the Corps never bothered to fill in the Delta end of the Mississippi River Gulf Outlet Canal. They simply left it to widen and turn more good marsh into open water.

"Success the Corps takes credit for; failure is an abandoned orphan," Turner chuckled. "I think your confusion is about the industrial end of MRGO. It served as a funnel during Katrina, allowing water to be driven straight toward New Orleans, like a hypodermic needle, as someone in the media put it. They've now plugged the city end of the canal. The idea is that in the next hurricane those immense gates can be closed and water supposedly will not reach New Orleans. I think the water may rush right over the gates in a real hurricane. It's a little like the geomorphology in Thailand, in which—"

"It's like New York," interrupted Jeremy, "The reason why the surge poured into Manhattan [during Hurricane Sandy] is because Long Island Sound is a funnel. Why do you think the tides are so high in the Bay of Fundy?"

"Why are the tides so high in the Bay of Fundy, Jeremy?" asked Turner, although he knew well enough, since he grew up in Connecticut, and sometimes lectured at Yale.

"Because, Gene, the Gulf of Maine is a giant funnel! The tides move faster in narrower confines, coming from the wider bay, and that's why you get a differential of fifty-three feet!"

"So, Caernarvon?"

"Caernarvon is a huge industrial structure somewhat similar in its look and scale to the near end of MRGO, but we saw it north of the city, and it is the state's main diversion project." Turner went on to describe the Caernarvon Freshwater Diversion project in detail. Built between 1988 and 1991 to "freshen" the intrusion of saltwater, Caernarvon is located near the Plaquemines–St. Bernard parish line. He explained that the $26.1 million cut in the levee has destroyed more marsh than it saved. His 2011 paper, "Freshwater River Diversions for Marsh Restoration in Louisiana," questioned the scientific basis of making freshwater diversions to restore marshland. He and his colleagues found that in three cases "long-running diversion projects . . . failed to increase vegetation cover or overall marsh area." They concluded, "Ultimately the scientific basis for river diversions needs to be more convincing before embarking on a strategy that may result in marshes even less able to survive hurricanes. The scientific basis for the benefits for freshwater diversions is obviously not settled."[18]

Gene reminded us of the greenery we had seen below the plane at Caernarvon. "It's green but it's only floating," he said. "It looked fine to us from the air since it is summer, but if you try to walk on it, you will sink."

It was time to say goodbye to Gene Turner. We would see him soon enough, when he trucked an LSU boat down from Baton Rouge to show us up close the varieties of Louisiana swamp, or why a marsh is sometimes not a marsh at all, and answer many questions raised by our aerial introduction. We would bivouac with Turner at a place called Woodland Plantation, sixty miles south of New Orleans, almost at the end of the Bird's Foot Delta.

What Is a Marsh Anyway?

Back together, we set out from Woodland Plantation in Gene Turner's little boat for his highly informative and sometimes amusing personal tutorial on the basic ecology of marshes and what makes them healthy or sick. In the near distance is Grand Bayou Village, near Port Sulphur (Plaquemines Parish) on the Bird's Foot Delta below New Orleans. The houses are mostly new, built on stilts—old homes destroyed by Katrina and rebuilt.

Turner jumps. He jumps high for a man in his early sixties. Each time his sneakers touch down, the marsh gives way, but holds together and springs back like a sponge. Turner is no stranger to this demonstration. Stepping to our boat, tucked in at the edge of the cordgrass, he takes out a heavy steel cylinder about two feet long with a capped end. This corer looks like post hole digger, what ranchers use to set metal fence posts into the ground. He holds it chest high, pushes it slowly into the marsh, wiggles it in a bit, then twists back, extracting a thick plug. Cordgrass is coarse and unruly, especially in its roots, which are what count to Turner. He cradles the plug in his hands, picking apart the dark wet soil and intricately entwined roots with his fingers. The tangle keeps the dirt together, like the robust roots of native prairie grasses back in Iowa.

Turner tucks the plug back into its marsh hole, ever the gardener. "Roots hold the marsh together. That's the first take-home," he says.

If the argument about how to restore Louisiana's marshes begins with water quality, one of the endpoints is this bundle of roots. Would they be even more robust (thicker, longer, healthier) if water from the river—from diversion pumps and spouts—flowed across their foliage? Or less?

Then Turner takes a shovel and digs around the plughole. He digs a foot or so, then steps back. We stare as the little hole slowly fills with water from the marsh. We stick fingers down to taste. It's brackish, a bit salty, though not so much as seawater.

"Marsh is not muck. Second take-home," Turner declares, "this is not solid ground. It's a floating mat. It rests only on or in water. The roots form an organic platform, a living thing, fashioned from plant material, and if we bring that dark soil back to the lab, under a microscope we would see that it's being formed, munched on, and made healthy by zillions of microbes."

Water floods the plants daily and keeps them soggy. Kept damp, these roots and the microbes create a living marsh soil full of dense organic matter and very little inorganic dirt. "The State doesn't want to admit that it's not just sediment," he chuckles. "If you drill down into the marsh, it's organic material, plant matter. There's very little mineral matter in there, which is what silt or sediment consists of."

He elaborates on this fact. Mineral matter makes up less than 4 percent of marsh soil at the coast. This has been the case for hundreds of years and the proportion actually declines as you go farther inland. When water loaded with nutrients floods these marshes, organic matter decomposes, and the all-important roots that hold the mat together shrivel and die. This means that nitrate-rich water destroys these marshlands. Flooding them with Mississippi River water is the worst thing you can do for them.

"And yet our whole management plan is based on sediment," Turner marvels. "The state built their approach, their analysis, their expensive solutions all on a geology model when this is living organic marsh, a biological system, calling for a different remedy." Mineral soil would be different. There, sediment deposition would add matter without eroding soil structure. But with soil that is almost entirely organic, "you add the nutrients and it chews up the carbon . . . then it gets weaker . . . then a hurricane comes and then it's gone!"

Turner motors us up the canal slowly but deliberately. He knows that snags and pieces of old pipe lurk beneath the surface because he's cruised here with graduate students many times before. The marsh soon becomes broken, and slightly higher. A network of banks supports some bushes and small trees. He runs us up on the edge, over dead oysters. "Now I want to show you a spill bank—"

"What kind of flies are these?" interrupts Jeremy.

"Those are called biting flies," Turner deadpans.

"I know that!"

He is worried that many insects have not yet recovered from the BP oil spill. Neither have many spiders. One of Turner's colleagues tested the health of the marsh by putting out traps with crickets in them. Insects breathe through holes in their hard outer skin called tracheal tubes. Four years after the spill, the fumes emanating from the oiled marshes were still killing crickets.

"How come there are so few birds?" Steve asks.

Turner explained the food web after the BP spill: "If there are no insects for the birds to eat, then, no birds, and the ones that are here have oil in their feathers, not because they were oiled in the BP spill and survived, but because the insects that are still here have oil in their own bodies and that's what the birds eat."

His hand sweeps across the Delta with its variegated grasses un-dulating in a soft breeze from the Gulf, glowing warmly in the early morning light. He explained, "They thought they had to get the oil out of the marsh, which came in a sort of puree we call 'diarrhea.'" Hundreds of cleaners hired by BP pulled their boats on top of the grass, galumphed around with their boots on, manually cleaning the oil off leaves and stems. In many cases this was worse than the spill itself. Turner compares it to remediation efforts undertaken after the *Exxon Valdez* spill in Alaska, where shoreline rocks were steam-cleaned, a ridiculously ill-conceived remedy that killed all the little things that bigger things ate. The whole system eventually starved.[19]

Jeremy explains that the same thing had happened in the Carib-bean in 1987. When oil tanks failed in Bahía las Minas near the Pan-ama Canal, oil flowed into the bay and the starfish all died because starfish also breathe through their skins. Ironically, the spill occurred next to one of the most studied coastal ecosystems in the world. Jer-emy was marine coordinator at the Smithsonian Tropical Research Institute, where most of this research had taken place. With his col-leagues, he investigated the changes that occurred, wrote the offi-cial reports, and published their results. They detailed for the first time what happened to a tropical coastal ecosystem contaminated by oil.[20] "People tell me that our report was the reason there was no oil drilling off the West Coast of Florida," Jeremy added.

Turner leaned forward. "What did you discover—because the Galeta Bahía las Minas spill was on tropical gulf waters fairly simi-lar to the BP spill . . ."

". . . and like any accident they might still happen off the west coast of Florida." Jeremy finished his sentence.

"What was the main take-home?" Turner wanted to know.

"That in a biogenic environment, like the marsh we see around us here or a reef like Galeta in Panama, or in a mangrove swamp, too, the entire structure of the environment is built by plants and ani-mals. If you kill those creatures, you kill the environment too, and it all turns to mud, and the mud then erodes into the sea." Jeremy used to tell Panamanians, "Panama is a smaller country now because of this spill. Part of your country has disappeared." And the spill hadn't gone away. The oil had settled into the burrows of crabs and other creatures. Without oxygen, it remained toxic, and when the rains came—and the first thunderstorm of the season in Panama can bring four or five inches of rain—oil would wash back out and it was like a new spill. You could smell the stink. "That's the second point: this stuff takes a long time to get out of the marsh. A freaking long time!"

We climb to the top of the pill-shaped spoil bank, chatting all the way. "Spoil" is the wet dirt and marsh piled up when a canal is dug. Until the 1970s, bucket dredges pulled the marsh apart, making long channels to bring in exploratory drilling equipment and, after a strike, to lay down pipe and otherwise move the gas or oil out to pumping stations, barges, holding tanks, and refineries. The number of canals dredged since the 1920s is in the thousands.[21]

The marsh and muck that did not wash to the Gulf during the years of dredging were piled into spoil banks. Soon the edges of the banks spread out and the cuts turned into open water. As the number of canals increased, so did the expanse of open water. "There is a one-to-one correlation between the area of canals and extent of land loss," explains Turner.

At the time, spoil banks seemed like a good thing. They gave the appearance of new land being created. Hunters built brush fires at one end to drive rabbits and the occasional deer toward their guns. Landowners used them to demarcate property lines. Oil and gas companies preferred spoil banks to the cost of filling in the canals, as coal companies must do in Appalachia or bauxite miners must do in Jamaica when they close their mines.

"Poor Jamaica!" shouts Jeremy. "Even Jamaica does it right! They make companies fill in their pits and restore it as useful farmland."

Filling in canals constitutes an external cost of doing business, which adds nothing to the bottom line. Today, filling a single canal can cost fifty thousand dollars or more, though, insists Turner, filling in the canals would be a low-tech, relatively inexpensive solution— especially compared with pumps, diversions, and wholesale dredging. A start to a solution.

Spoil banks create a second problem because they often change the natural north-south flow of water to the Gulf, and stop the flow entirely at intersecting elbows and corners. The still water forms ponds, which disintegrate the banks further and create more open water. There is a bureaucratic driver to all this as well. In Louisiana, once private land turns to open water, it becomes the property of the state, conferring all-important mineral rights away from the private owner. So it becomes a land grab.

Turner takes out his shovel and shows us that the older soil of this spoil bank is sandy, dredged from the bottom—what he calls mineral soil. When a canal is dredged, the bottom is heaped on top of the marsh, killing it. Turner has done the calculations. For every canal dug, six times that area of marsh is eventually lost, a startling multiplier that explains why marsh is dissolving.

Close to the Gulf, the artificial ditches created by spoil banks run at right angles to the coast. This lets saltwater flow north into the marsh, changing its salinity, and further drowning and killing the plants. Farther north, spoil banks tend to ring the marsh, allowing nutrient-rich (not so fresh) water to drown it, Turner says, much "like leaving the sprinkler on too long in your front lawn." In dry seasons, the spoil banks desiccate the marsh, oxidizing the organic matter in the roots. This kills the marsh from the opposite direction.

As if to punctuate his point, Turner jumps again, this time on the top of the spoil bank. He lands with a thud. He smiles. Point made. We motor back.

Queen of the Dead Zone

Nancy Rabalais swims down through mucousy strands of algae and bacteria. She wears a full wetsuit under SCUBA tanks. Visibility is minimal in the Gulf of Mexico this day. The research vessel *Pelican* rocks above her head. The wetsuit is as much to protect her from chemicals and bad water as from the cold. Most of her collection efforts are not so personally involving. Usually, various bottles and sensor devices are winched off the *Pelican* in the annual one-week,

1,200-mile transect off the coasts of Texas and Louisiana to measure the extent of the Gulf's "dead zone" of oxygen-depleted waters.

Rabalais, whose sobriquet "Queen of the Dead Zone" cuts several ways, was a 2012 MacArthur "genius award" Fellow. Scientists and artists are given this award for thinking out of the box, often stirring up the waters, as Rabalais has certainly done throughout her life. Besides her more scientific samples and measurements, Rabalais each year collects a bottle of water from the dead zone and pours it into an urn at the Unitarian Church near her home in Baton Rouge, "praying," if that's what earth scientists do, for cleaner seas and better water for our country. She applied most of her $500,000 grant toward filling gaps in her own research funding and supporting additional research by graduate students, keeping little of the money herself.

Until July 1, 2016, Rabalais was director of the Louisiana Universities Marine Consortium—LUMCON—headquartered about a thousand miles due south of Dysart, Iowa, in the tiny fishing village of Cocodrie on the Gulf Coast between the mouths of the Mississippi and the Atchafalaya. LUMCON floats off the bayou like a space center: it is an ultramodern building with lots of glass, painted a strangely festive puce. There is an observation tower on top and tall cement pilings below so that research boats can dock at high tide.

Rabalais, sixty-six, describes herself with a raucous south Texas laugh as "oil field trash and proud of it!"[22] Born in Wichita Falls, Texas, she grew up in the drilling grounds of Odessa, then Houston, and finally Corpus Christi. Her father was an engineer for various oil companies and then for Reynolds Metals. She worked her way through three degrees.

When Rabalais first came to LUMCON in 1983, she had just earned her Ph.D. from the University of Texas, "chasing fiddler crabs around on the beaches of south Texas." Female marine biologists were an unfamiliar sight in the region at the time, and guards often accompanied them on nighttime collecting, to keep them safe.

In 2003, after LUMCON had been without a director for four years, Rabalais volunteered her services. "We're suffering," she told her colleagues. "We need a director and I think I would be a good one. I don't know how many times I said that before they started taking me seriously. I don't think it was a woman thing. I'm not sure. I don't really think so."

Nancy's office overlooks the marsh, which is cut up by channels once used for oil and gas exploration. Some of the channels have widened to open water. The patchwork pattern of canals and marsh stretches down to the Gulf. Her office walls are lined with charts and marine art, which she collects, including a brightly painted wood carving of a medusa jellyfish and rare prints of flying seabirds inked by the iconic Mississippi painter Walter Inglis Anderson.

What's pulled Rabalais out of the reeds and backwaters of south Louisiana and put her before Congressional committees is her work on the Gulf's dead zone. This makes her famous in some quarters and infamous in others. Although she didn't invent the term, she's been saddled with the vampiresque moniker "Queen of the Dead Zone" ever since she and her husband, Gene Turner, published the game-changing paper "Coastal Eutrophication Near the Mississippi River Delta," which made the cover of the journal *Nature* under the catchier title "Mississippi River Delta Blues."[23]

A dead zone is an area of water so low in oxygen that most sea creatures must swim out of it or die. The dead zone in the Gulf of Mexico occurs when agricultural chemicals, mainly nitrates, but to a lesser degree, phosphates, wash out of the Mississippi. It is among the world's largest dead zones, along with China's Yangtze and Europe's Baltic. Scientists correctly predicted that it would be the largest ever measured in 2017, when it covered 8,776 square miles.[24] The fertilizers cause tiny floating algae and other microorganisms collectively called phytoplankton to bloom explosively in unsustainable numbers, just like the deadly cyanobacterial blooms in Lake Erie's dead zone up north. The principle is no different from fertilizing your front lawn, or pansies on the kitchen window, although draining the lawn into a frog pond to create a burst of scum would be a more logical comparison.

Tiny grazers called zooplankton eat some of the proliferating algae, and their even tinier fecal pellets wend their way to the bottom. Other algal cells simply die and drift slowly down. Some of the dead cells form aggregates that look like mucus as they sink to the bottom. Nancy elaborates: "You can see it when you're diving in it, streamers of junk falling through the water. And once that gets to the bottom, bacteria start to break down that carbon and consume oxygen at an incredibly high rate 'cause so much carbon is getting to the bottom[, and] at a faster rate than the oxygen can naturally get from the surface to the bottom." In other words, there is so much food for the bacteria to munch on that they use up all the oxygen; there's none left for the animals near the seafloor to breathe. This phenomenon whereby almost all the oxygen in seawater is removed is called hypoxia (or anoxia if every last molecule of oxygen is gone). The hypoxic layer can extend thirty feet from the bottom in a sixty-foot water column. Fish and crabs that can't get out of the way float to the surface and die.

Further, "the river brings fresh water, which sits over the denser salt water so that you have two layers." The density difference between those two layers creates a barrier called a pycnocline, across which waters do not mix well. "In the summer the sun warms the surface while the bottom waters stay colder, like in a lake in August where you can have warm water around your shoulders and your feet are cold." It's just simple physics. With the density barrier in place, oxygen from the air—and from phytoplankton during photosynthesis—doesn't diffuse downward to the bottom waters with lower oxygen. Although the layers get mixed up in winter and during hurricanes, areas can become depleted of oxygen, or hypoxic. "This physical system's been here forever, but hypoxia, or oxygen-minimum zones, have not. What's changed over time is the excess nitrogen and phosphorus getting into the Gulf and that's mostly happened since the fifties."

Shrimpers first reported the Mississippi dead zone in the early 1970s. The waters off the mouth of the Mississippi had always been rich in the nutrients that supported Louisiana's incredibly rich fisheries.[25] Larval fish eat the phytoplankton and zooplankton, and they get bigger and move up the food chain. This is why the area around the Mississippi was called the Fertile Fisheries Crescent. But not anymore. Instead of a fertile crescent, there is a huge area of low oxygen caused by all the excess nutrients from farming way up north.

What joins Cocodrie and the Gulf of Mexico to Dysart and the lakes, ponds, and groundwater of America's Heartland, then, is too much of a good thing in the wrong place.

Mississippi's dead zone has averaged a little over 5,500 square miles over the 2011 to 2015 period, although during drought conditions in the Heartland it can fall to 1,700 square miles. Rabalais has been measuring it aboard LUMCON's *Pelican* since 1985. In 2015 it was the size of Connecticut and Rhode Island combined—6,474 square miles—the eleventh largest area since measurements began, and in 2017, as mentioned earlier, it broke the all-time record at 8,776 square miles. That's more than four times larger than the purported environmental target of 2,000 square miles.

"We used to say it covered an area the size of New Jersey," she once told an audience in Washington, D.C., "but Christine Todd Whitman, the former governor of New Jersey and at the time the head of the Environmental Protection Agency, came up after my talk, and said, 'Why do you always have to compare that thing to New Jersey?'"

After President Richard Nixon signed the EPA into law in 1970, environmental impact statements became necessary before offshore drilling could begin. Routine data from an early Texas A&M research trip funded by an oil company contained one puzzler: a measurement of low oxygen in the waters off Freeport, Louisiana. Back-to-back floods had hit the Midwest in 1972–1973, and immense amounts of farm discharge had passed down the Mississippi and into the Gulf. Examining the records separately, Nancy Rabalais and a young researcher named Gene Turner (they married later) realized that the periods of depleted oxygen had occurred in summer months, after the spring floods in the Heartland. They reasoned that excess fertilizer off the fields, especially the big row-cropped fields of corn and soybeans, had drained off in the tiles with the water. Nitrate and phosphorus were already causing problems in Midwestern lakes, but nobody had thought to check the Gulf of Mexico. Nancy shrugs. "Nobody believed us when we first said the increase in fertilizer use was coincident in time and space with the worsening hypoxia." Their *Nature* paper offered the first hard evidence that the Gulf's dead zone was caused by the common practice of replacing natural fertilizers with anhydrous ammonia—the highly productive way that Americans had begun to farm after World War II.

"Corn grows better if it's not sitting in soggy water," Rabalais explains, echoing John Weber. Starting in the 1920s, marshes across the interior of the United States were ditched and drained to make cropland. Now the entire landscape has a pipe under it, and nitrogen dissolved in groundwater comes through this drainage. A map on her office wall, *The Nutrient 9*, shows the nine states currently most responsible for the nitrate coming down the Mississippi and entering the Gulf. Iowa and Illinois lead the (dis)charge.

Biologically bound silica (BSi) was key to Turner and Rabelais's argument in *Nature*. Diatoms are elegant, one-celled algae that form silica, or glass shells (when the paper came out, a highly magnified diatom graced the journal cover). Deposits of diatoms on the floor of the Gulf of Mexico can point the way to oil deposits because oil is believed to come from vast deposits of sunken algae pocketed today under ancient sea beds. Incredibly important to the survival of the oceans and all life on earth, these creatures take the sun's energy, turn it into sugars, and release oxygen into the atmosphere during photosynthesis. Large and small algae account for 45 percent of primary (sugar) production, and for at least half of the oxygen on earth (by comparison, terrestrial rainforests provide about 20 percent of the earth's oxygen).[26] You could say the lungs of the planet are wet.

What makes diatoms invaluable for scientists is that, after diatoms die, sink to the bottom, and decompose, their shells remain in the sediment as BSi, which fossilizes and can be measured. Turner and Rabalais showed that, over time, more and more diatoms had accreted into the sediments on the floor of the Gulf of Mexico, proof that the primary productivity in the water above was getting higher. Algal blooms forming on the surface were being fertilized by the increased nutrients.

Turner studied cores drilled into the ocean floor to discover which years had the most diatoms. Reading cores is like turning pages in the book of ocean time. It provides forensic evidence for what happened. The cores showed that BSi accumulation rates increased as much as 100 percent annually during the twentieth century, and were greatest under water 65 to 165 feet deep, less than 65 miles from the Mississippi outflow. Nitrogen in the river also doubled between 1900 and 1980. An inexorable rise of diatom deposition from 1950 to the present overlapped nicely with rising rates of nitrate flowing into the Gulf.[27] Today, the best way to predict the size of the summer dead zone is to measure the nitrate in the Mississippi below Baton Rouge, as Turner does.

Nancy and Gene subsequently published another paper, "Sediments Tell the History of Eutrophication and Hypoxia in the Northern Gulf of Mexico," in *Ecological Applications*, and this time with Rabalais as first author.[28] The paper showed that nutrient overload began two hundred years ago with the clearing of old-growth forests from Minnesota down through Louisiana, just as carbon dioxide in the earth's atmosphere can be shown to have risen steeply after the Industrial Revolution.

Rabalais and Turner took cores in areas where sediment had piled up, and dated the layers of sediment using the signatures of lead-210, a radioactive isotope that decays at a steady rate over time. Thus they made a time series of core layers that went back to the 1800s, which allowed them to determine when the diatoms had died and fallen to the bottom. Rabalais found, too, that pigments of the phytoplankton correlated with anoxia because diatom pigments last longer under anoxic conditions where no oxidation takes place. Barun Sen Gupta, a geologist colleague and co-author on the paper, showed that different species of Foraminifera (microscopic protozoans with a calcium carbonate shell) actually preferred low-oxygen conditions in which other species died off. He gauged worsening oxygen conditions by measuring ratios of species concentrations.[29]

Rabalais and Turner even assayed the radioactive isotopes contained in plants and soils. It is possible from these isotopes to distinguish the sources of nitrogen in the different Midwestern states, and even to pinpoint which counties within the states released the most nitrogen. Nancy explains: "We looked at stable carbon isotopes, and if it's a land-derived carbon, it has a different signature than diatom-derived carbon. Terrestrial carbon is all near the shore and right around the delta. High biogenic silica signals near the river come from freshwater diatoms. They have ten times more silica per body volume than marine diatoms, but they fall off really quickly. The other carbon is from phytoplankton growth, which means it's been stimulated by the river." So neither carbon from the river nor freshwater diatoms provide the carbon in the dead zone: it comes from marine phytoplankton. Nutrients stimulate the phytoplankton to grow too fast and grow too much. Then when their carcasses sink to the bottom and rot, the bacteria feeding on the debris use up all the oxygen faster than it can be replenished from the surface. This creates the dead zone.

"How do we write a paper?" Gene Turner laughs across their breakfast table. "I write a sentence. Nancy corrects it. I write a sentence. Nancy corrects it. I . . ."

"Well," drawls Nancy, "That's on the papers you write."

Rabalais and Turner had nailed it from a scientific point of view. Still, many agricultural scientists working within and outside industry adhere to the idea that "natural variability" accounts for the seasonal rise and fall of nitrates, of algal blooms, and the extent of the dead zone. "The natural variability thing is what the Dark Side of the Force will say about anything until you prove otherwise," Nancy sighs.

Sixty miles south of New Orleans, near the end of the Bird's Foot Delta, is the welcoming refuge known as Woodland Plantation. Originally built in 1834, and formerly featured on the label of Southern Comfort—the South's iconic, spiced bourbon—it is now an upscale, somewhat eccentric fishing lodge complete with a gothic church that serves as a restaurant and bar.[1] Turner had dropped us off there before he headed back to Baton Rouge. He stays there often on his Delta research trips, and told us we would be "quite comfortable." That was quite an understatement. Not only was it comfortable, it was so central to the workings of the deep Delta that it became our improbable southern field headquarters.

As we walked into Spirits Hall, the cavernous church that serves as the lodge's dining hall, a massive American Staffordshire terrier stood upright against the bar as a young woman lobbed ice cubes to him over her shoulder from the sink.

The dog, Frank, caught the ice cubes, snapping his huge head to the left and to the right with a crack and a crack and a crunch. It was mesmerizing and a little scary. "Oh, don't y'all mind Frank," Foster Creppel, the proprietor of Spirits Hall and Woodland Plantation, greeted us. "He's just a big baby. Frank! Get on down!"[2] Dappled Frank is Foster's dog, and he is enormous. West Coast pit bulls are briefcase models of Frank.

"What are the stitches for, on the back of Frank's head?" Steve asked.

"Oh, Frank," sighed Creppel, "he got in a bit of an altercation with my neighbor's dog, and he couldn't be beaten off with a shovel handle. There were some small children on the porch—I get along fine with my neighbor—so my neighbor had to shoot him a couple times. Didn't bother Frank much, but it did get his attention. I drove over as fast as I could, and when I got there, I think my yelling hurt his feelings more than the bullets, which were just .22 shorts. One bullet creased his left testicle and the other went clean into his head. He's a very stubborn animal. A very lucky animal. I'm thinking of changing his name to Lucky Frank. Frank! I said get down!"

Creppel had hauled Spirits Hall, an 1880s Gothic Revival church, onto the grounds of Woodland Plantation to serve as a restaurant and gathering place. "Spirits of Woodland Hall," as it was formally christened, has oaken floors that are leveled, dining tables where the pews once stood, soaring stained-glass windows, and a baroque mahogany bar in place of the original altar. Rows of alcoholic spirits rise behind the altar bar where ministers once held forth, and the higher the shelf the pricier the offering: from Captain Morgan and white Malibu Caribbean Rum on the bottom, to Knob Creek bourbon, and on up to Macallan's single malt Scotch. The very pinnacle is occupied by a bottle of Southern Comfort. Southern Comfort was king of the roost because a romantic lithograph of Woodland Plantation by Currier and Ives, called "A Home on the Mississippi," was on the popular label from 1934 to 2009.[3] The distillery still holds retreats and contests here.

Woodland Plantation is on the west bank of the Mississippi at Pointe à la Hache, about thirty-one miles from the Gulf. Built by Captain William Johnson, an early river pilot in 1834, it once anchored a sugar plantation of 2,100 acres and about 210 mostly African slaves.[4]

Despite its ambiguous history, Spirits Hall soon became our Delta field headquarters, an ever-enlightening and tasty jumping-off point for two road-weary adventurers from the oft-maligned worlds of science and journalism. Spirits Hall ministered to the heathen and the believer alike, as well as the secular and the zealot, climate-change researchers and climate-change deniers—Americans all.

Spirits Hall offered martinis and gin, Southern comfort of all persuasions, good local beer, and fresh lemonade. Nightly conversations spoke to the controversies swirling around water, that ubiquitous liquid that in south Louisiana meant boon or bane to conservationists and sportsmen, oyster and shrimp harvesters, ecologists, watermen, engineers, oil and gas executives, and ordinary residents. Many local representatives of these groups had gone to high school with Foster Creppel. They had partied together in the French Quarter, or yelled at each other from passing watercraft, sometimes for decades. All gathered companionably each night after cocktails had been poured, the shrimp shelled, and speckled trout and redfish cleaned—displayed before us *in situ* not long after being *in vivo*. (A science joke meaning the food is fresh here.)

We felt like Darwin landing at Santa Cruz Island in the Galápagos, or maybe just a little like Frank: lucky. Turner had delivered us to the right place at the right time, not too long after Katrina, Gustav, and Isaac; just after the BP spill and cleanup, during the controversies between the levee board and the oil company lawsuits; and as Hurricane Sandy was waking up the East Coast to the fact that what was happening to New Orleans was not just some distant tragedy, isolated aberration, or HBO show.[5] Maybe New Yorkers would be next.

Our task, which we accepted, was to parse it all out.

"I like to walk my dogs," Creppel said after we had agreed to stay the night. "You guys want to come along?" An hour before dinner—time for Lucky Frank's constitutional.

We walked out the back of the hall, where ministers had once chatted with parishioners after church, down the steps, across the lawn, and along the winding gravel road past the alligator ponds to the back levee along the Mississippi. As we strode along, huge ships from China moved slowly upstream to New Orleans.

For his part, Frank kept running down to the ponds to scare up mallards and coots, and to worry gators that peered up with yellow-black eyes from the watercress and moss until, at the last second of his approach, they slipped under the filigree of green algae covering the surface. At times Frank would spy a woodchuck or a squirrel (or imagine he had) and bound into the jumbles of gray granite riprap trucked down from Kentucky or Tennessee by the Army Corps of Engineers in the hope of holding back one flood or another.

Foster Creppel is a former corporate recruiter turned innkeeper. He walked fast, not so fast as Frank, but he kept up a good clip, talking all the while. Creppel's father had been a fighter pilot and his parents owned The Columns, a hotel in New Orleans' Garden District, but Foster had a love affair with the Delta. "When I was nine or ten, I used to hunt along Bayou Barataria, which was like a smaller Mississippi River then, a big distributary, and I'd ask my grandmother, I'd say, 'Why do we have all these oaks and palmettos and black squirrel on this side and right across Bayou Barataria it's swamp and marsh? And she'd say, 'Well, this is because Bayou des Familles ran off the Mississippi River and into the Bayou Barataria and created a ridge, a high ridge.' That ridge all along that is where they developed Jean Lafitte National Park.[6] Tributaries build rivers, distributaries build bayous."

Creppel believes that water will restore the Delta in the way that flowing water created Lafitte National Park. What the system needs is good water: silt-laden, café-au-lait brown from the Big Muddy and all its tributaries up north—the Missouri, the Ohio, and maybe Wolf Creek, which drains John Weber's fields in Dysart.

He is a proponent of turning the Mississippi loose again, of freeing silt from the huge upriver impoundments behind dams in North Dakota, Montana, and Minnesota, and letting its water fan out and be filtered once more by a restored marsh. In his view, this would clean up the dead zone, too. He has strong opinions. "All that sediment should be moving! . . . I don't understand how the Army Corps can flood the upper Mississippi and let all that water run off, while down here it could spread out and cover parts of the Delta and restore it the way it used to be, and yet, here, they let the water rush out into the Gulf. I ask that question a lot, and I don't get a good answer! Fresh water built this delta in the first place. It'll take forty, fifty, maybe two hundred years, but that's what we need to do. You can't put back together in four or five years what it took a thousand to build. We engineered the Delta's destruction. I think we can re-engineer its success and rebirth." He runs an eye down a line of tall poplars flanking the river. "Frank, get back here!" he shouts.

"See there," Creppel points down the levee, past the trees into the sandy brown water. "That's where we've lobbied the state to put a suction dredge and a siphon and suck a lot of mud up and pump it about two-and-a-half miles south, under the road, past Jefferson Lake, onto Lake Hermitage, to rebuild the shoreline." Under the Coastal Protection and Restoration Authority (CPRA), in 2012 the sediment started to be pumped to Lake Hermitage from borrow sites on the Mississippi River. In August 2014, however, observers at Lake Hermitage (and another site nearby, Bayou Dupont) noticed that mixed in with the deposited sediment were coal and petroleum tar debris covering about a quarter of the 2.5 acres of restored ground. While the CPRA maintains that no problems will likely result from this, the source of the debris, United Bulk Terminal, consented to give $16,500 to the Woodlands Conservancy for improving the Trail and Bird Park as partial remediation.[7]

"Frank! Get out of there! He'll roll in anything nasty." Creppel continues, "It would be good to fill in some of these oil and gas canals, too! We will need closer to half a billion every year and for many years. Mineral sediment is heavy and organic is light and floats. They are equally important to grow the Delta. The chenieres [marsh ridges of shells or sandy dirt usually with trees on top] or trees grow on both. It's a living system—coontail grass, willows, widgeon grass, bugs, they are all part of blackwater swamps. Deltas are built from the river out, from the inside out!"

It's a punctuated monologue, and he pauses for breath.

"But it would take a lot of education, and political and local will, and people need to understand that it's not going to kill their way of life if we have to reintroduce fresh water . . . You can't restore the estuary without fresh water. There were a lot of freshwater swamps and marshes in this part of the Delta. That's why we had ducks and mink and muskrat. We don't have any ducks, mink, and muskrat anymore, because they need fresh water! Frank, come here! Damn that dog is hard-headed."

Not everyone agrees with Creppel's vision of restoring the Delta with freshwater and silt. In fact, this is the crux of the argument about how to save the Delta (or at least one part of the argument) not just in Louisiana but in coastal Mississippi and Texas, too.

How would the Delta get more freshwater? And is too much river water a bad thing? Does the water flowing down the Mississippi these days possess the restorative properties it once did, as Creppel fervently believes, or does it instead weaken the marsh, causing the roots of the saltmarsh cordgrass to remain short, as Turner can demonstrate? From a scientific point of view are Foster Creppel's artful monologues mostly wishful thinking, even if they reflect some common sense?

We soon heard other opinions. Oyster growers, for instance, believe that freshwater will kill the oyster reefs, which require precise salinity from a mixture of freshwater and saltwater. Too much salt, for instance, brings in predatory boring snails, or oyster drills that feed on shellfish. Shrimp need the marsh to incubate their larvae, and shrimp fishermen believe that too much freshwater would kill the larvae and so their business. The different species of shrimp (brown, red, and white) figure into the equation, too.

Oil and gas companies are determined not to be blamed for having eviscerated the Delta starting in the tidewater oil-boom days of the 1920s, when they began digging ten thousand miles of exploratory canals—work that continues unabated today. They especially do not want to be forced to fill in all of these now much widened canals, which might be an easy, cheap, low-tech first step to rebuilding the shredded marshland. Most of the people we talked to—the scientists, Creppel, and the oyster folks—believe this might help. But that would be very expensive for the oil companies. The state of Louisiana, along with sand and gravel interests, prefer the vastly more expensive solution of rebuilding the barrier islands offshore, dredging up sand to pump onto sections of the marsh, and building more levees.

Leading marsh scientists, including Gene Turner, point out that the silt-laden water Foster Creppel puts his faith in is not the same as it was before 1951, and is very different from two centuries earlier, when the forests of the upper Mississippi had not yet been cleared.

The river has been channelized by bold engineers and pilots like Captain Johnson of Woodland so that it will not meander over the Delta, but instead will rush out into the Deep Gulf. Up north, the Mississippi and its tributaries have been impounded in big dams and reservoirs that were created by Franklin Roosevelt during the Depression, and in many cases displaced beleaguered Native American communities.

Of all the countervailing interests and arguments, this may be the hardest for Creppel and others to swallow: the nitrate- and phosphate-laden water that makes it downriver is not much good because it induces decomposition and causes marsh plants to send out shorter roots. This process weakens the marsh and, ironically, makes the Delta even more vulnerable to hurricanes. In fact, a strong hurricane can roll up miles of poorly rooted grass in what is described as an accordion affect. The marsh crunches in long messy folds that are more easily ripped out and blown away—like a wind-driven accordion.

But what if these elements of the ecological puzzle could be fitted together again in some sort of working system? This would mean deconstructing Heartland dams, with the millions of tons of silt filling and clogging them released downriver; taking down levees so that the silted, flowing water could spread over marshes and fields; raising houses and roads to allow for this to happen in human communities; and somehow persuading the U.S. corn and soy industries to fertilize less or with pinpoint precision—because as Weber points out, these unrecoverable "inputs are expensive." What if, even with all these measures, the whole coast continues to sink, as it will, and the waters of the Gulf continue to rise, as they will? Would all this remediation be for naught? Ten, twenty, or thirty years from now, will Baton Rouge become a waterfront city, with New Orleans reduced to a human-built fill-island or a low-lying country like the Maldives? Or might the Big Easy suffer the indignity of serving only as a forward base for oil, gas, and navigation facilities, with perhaps a leveed and tamed French Quarter reduced to entertaining workers jet-coptered in from Houston? Is that what the future holds for this Cajun-style, human-built world?

Put more succinctly, the whole region—everything everyone wants to save—may one day be underwater because of natural subsidence, sea level rise, and more powerful storms.

Food for thought. Time for dinner.

"Frank, get over here!" shouts Foster.

Time in Plaquemines Parish is told in terms of hurricanes endured, as in, "See that fishing boat in the borrow pit? Katrina put her there. Nobody's claimed it yet." Or, "It was mostly Ike that blew down the trees on the cheniere." Or to listen to Amanda Adolph, the breakfast hostess at Spirits Hall, "When the wall of water came over the back levee during Isaac, I only had a few seconds to run 'cross open ground to the Big House. I closed the door and the water smashed against it. I called my mom in her trailer and they took me out with an airboat from the second floor."[8]

Hurricane Katrina, the second costliest national disaster in American history, hit hard in Plaquemines Parish. Katrina is the defining twentieth-century hurricane in Louisiana, just as the flood of 1927 was the flood to measure all others by.[9] The majority of structures still standing, even now, belong to three categories: oil and gas facilities, navigation centers, and FEMA trailers. The FEMA trailers all look as if they could blow away in the next hurricane, while the energy facilities and the navigation centers could have been built by the sensible pig in the children's story, all engineered brick and steel, most fifteen feet off the ground with pillared open space for the water to rush under.

But although Katrina wreaked Category 5 havoc on the parish during its short run to New Orleans, Hurricane Isaac's lower winds and higher waters were actually worse for Woodland Plantation, the only other structure of note to survive in these parts. In the aftermath of Isaac, alligators in Woodland's freshwater ponds drowned with the infusion of saltwater. Cows drowned, too, and died of thirst and starvation since they couldn't eat the salty grass. The grass itself soon died from the excess salinity, along with rows of blood orange trees raised by Woodland's neighbors.

"It was a mess, I'll tell you that!" laughs Creppel. "See y'all tomorrow!"

Breach

In the morning, after a light Plaquemines breakfast of poached eggs, fried potatoes, strip bacon, hominy grits, fresh bread and homemade biscuits, stoneground oatmeal, black oyster gravy, fried spinach, blood oranges from Foster Creppel's neighbor's trees, freshly squeezed orange juice, rounds of Louisiana roasted coffee, and a little strip steak or grouper (if you want it), we drove north to investigate the scientific basis of Creppel's strong views that fresh deposits of river water could help to rebuild the Delta.

In the Winn-Dixie parking lot on Paris Road in Chalmette, we met with John Lopez, his Ph.D. assistant Ezra Boyd, and his boatman and colleague Andy Baker. Lopez, coastal director for the Lake Pontchartrain Basin Foundation, is a short, full-bearded man who that day was wearing a clean orange expedition shirt and blue jeans. Back in 2005, Katrina had ravaged his home at the eastern end of the lake.

For ten years, Lopez had been measuring what he called "overtopping" with soil cores analyzed at the Scripps Institution of Oceanography. Ratios of silt to sand and silica told a tale of high and low water, flooding and drought. He wanted to know how much water was flowing off the main Mississippi over time through the Bohemia Spillway (what used to be called a breach or crevasse), and how much soil the river was depositing naturally.

The Bohemia Spillway has a backstory. It was not a natural break in the levee but an intentional breach. It had been exploded, thereby creating an *in situ* experiment. Ironically, the Bohemia Spillway is, perhaps, the one place where the Mississippi River has been allowed to flow naturally, at least in part, for over ninety years.

During the torrential flood of 1927, some wealthier citizens of New Orleans dynamited the levee, not to save the residents of New Orleans and those just downriver, but to salvage nearby sugar plantations and the banks that held the notes on them. As John Barry wrote in his definitive book *Rising Tide*: "Pool's bank was the most vulnerable in the city; he had aggressively loaned money to the sugar planters. A crevasse on the river's west bank could destroy them, and his bank. Dynamiting the levee on the east bank might also relieve them. Pool argued: The people of New Orleans are in such a panic that all who can do so are leaving the city. Thousands are leaving daily. Only dynamite will restore confidence . . . How real was the threat to New Orleans? The threat to its business was real enough, but how real was the threat of the river? Or did it matter?"[10]

The water kept rising. The levee was blown. Homes were destroyed, people became refugees, St. Bernard and Plaquemines parishes were devastated. But the hydraulic boil pressing on the city and certain leveraged plantations was lanced. There were further ironies, as there always are in Louisiana. The flooded land, believed to be worthless, turned out in the 1930s to have oil under it—the new black gold—including the Potash and Cox Bay oil and gas fields. So the trappers and truck farmers who had been flooded out of their worthless swamp wanted compensation at a vastly higher level. Suits and countersuits continued for decades, into the time of their grandchildren's grandchildren, until in 1984, the descendants won their lawsuit and much of the land was transferred back.[11]

Lopez drove us briskly through the city and south to his research site atop the breach, pulling behind his twenty-two-foot craft, the *Save Our Coast*, a bumper-sticker name for a river-worthy boat. Along the way we passed the "Great Wall of Chalmette," as one extensive levee is called, and mentally revisited Katrina, as everyone here always does, and will.

Along the way, Ezra Boyd spoke of his 275-page Ph.D. thesis, "Fatalities Due to Hurricane Katrina's Impacts in Louisiana," which featured a telling formula for just how rising waters and storm surge can kill in a hurricane: "Depth times velocity equals the killing power of water," Boyd explained in a gravelly voice. "Six inches moving fast—and we tried to plot the speeds near the dead bodies, although they can come to rest far away—can sweep you off your feet and you are gone, just as much as with slower, deeper water."[12]

There was a silence in the car. "An equation that explains death," Jeremy finally laughs, "That's fascinating! That's macabre."

But the day was sunny and beautiful and the river low. We launched the *Save Our Coast* at a slide ramp into the main river, then quickly disappeared up a nameless stream. As the depth decreased, we pushed ahead ever more slowly. Bushes and then trees began to cover the banks. Finally we tied up beside an old road used to bring in drilling equipment. Boyd warned us to look out for spiders, and there were plenty—two-inch black-and-yellow ones with nice scalloped markings—but they were pretty much harmless.

Lopez began to plumb the depth of the overflow with a marked measuring stick, calling out numbers. The drilling road had caved during a more recent flood, and he and Baker accessed the banks with an aluminum canoe marked with Native American designs that had been stowed in the bigger boat.

In the ninety-odd years since the powers-that-be in New Orleans had blown the levee to relieve pressure on the city, mineral soil had built up a few inches. This was momentous to those who felt that silt from the Mississippi might bring back solid ground, allowing bushes and trees to recolonize the area or be planted, to eventually become a line of defense—a natural earth bulwark—in front of New Orleans. The logic of the argument points to removing more levees; pumping or dredging soil from the main river to deposit in eroded areas, as at Lake Hermitage; removing spoil banks; and so on.

The big problem with this method concerns the amount of silt the river can still bring. The amount from 1927 to roughly 1937, when the Depression-era impoundments were mostly finished and the lower third of the river channelized, was considerably more than in succeeding years because of the dams and locks upstream.[13] And what sort of marsh was this? The ground was relatively solid—what Gene Turner and John Lopez refer to as mineral soil. This was not the spongy organic marsh found farther south, and therefore it was not so vulnerable to nitrogen and phosphorus overload. There had always been some elevation here at the Bohemia Spillway, forming a small natural levee, a minor but important sort of high ground first built on by early residents of New Orleans who had more respect for hurricanes than have newcomers over the last fifty years.

Lopez pointed out a crucial difference between the swamp we were plumbing and the spongy marsh closer to the Gulf: there are fewer oil and gas canals here, although you see some. Perhaps 20 percent of the land loss around Bohemia could be attributed to oil and gas canals, he told us, but the percentage was much higher outside the spillway. He believed that land loss near the Gulf because of oil and gas canals was 40 to 50 percent overall, but reached 80 percent in some places. And this did not account for future exploitation closer to the coast, where, said Lopez, some 80 percent of the land was privately owned. Owners had a legal right to cut canals and drill, and why wouldn't they? Oil and gas are the state's economic engine.

"That ain't going away," said Jeremy, "even if New Orleans does." Lopez did not respond to that.

"But it's not just oil and gas canals," he continued. "The whole coast was being industrialized at about the same time—navigation canals, shipping canals, the regular tracks from boats, even airboats, maintaining power lines in the marsh. The larger oil and gas canals, eighty to a hundred feet wide, show up on earlier maps but the smaller cuts do not. Satellite imagery did not begin until the 1950s and, ironically, the excellent older maps from 1932 to 1950 did not jibe with the lower resolution of the satellite imaging, so smaller canals were actually excluded from the later data sets. Spatial resolution was lowered to provide what I would consider to be a false continuity."[14]

"In the 2011 river flood," Lopez added, "water was going over the crest of the natural levee in the Bohemia Spillway, but then it was captured by channels, even in the natural drainage part, so what you saw was not widespread overland flow, but water flowing into the channels. We think the deposition is occurring in the channels or the little bays the channels flow into. The wetlands here in Bohemia have a low rate of loss, and canals have actually filled in, as opposed to canals elsewhere where the marsh breaks up. We think it is because of the connection to the river. Is the marsh healthier here?" He shrugs. "That's what we are measuring."

Some of the remaining river sediment flows out into bays in storms, as elevated water levels stir up the bottom and redeposit sediment from the marsh. Lopez summed up, "Wetlands are doing better here, canals are filling in, water is coming from the river, though we are still trying to see why. It is a re-creation of the natural processes. Before 1850 there were no levees, no people; 1850–1926 people built artificial levees; and in 1926 the natural conditions were re-created by removal of the artificial river levees. The river was re-engaged, and instead of a cascading wetland loss it is recovering with zero loss to slight gain, very small accretion, to be defined . . ."

"Interesting that there should be a source of sediment," says Jeremy as we sit on the bank of the little distributary, eating lunch in the sun, "A question: if this was before the dams had been built up-river, there would have been a lot more sediment, right?"

Lopez agrees. "We do know it is less. Sediment content went down after the dams were built. But the sediment layers were higher artificially from Midwest farming and stirring up.[15] Nobody really knows [the exact figures]. We look geologically at the Delta and the total contribution of the river and what goes out the mouth of the Mississippi and fills the Gulf of Mexico, then how much goes over the bank and fills the wetlands. Say 75 percent goes out, and only 25 percent goes to situations like Bohemia, the sediment is down and never captured by the marsh. So take out a disproportionate amount of sediment from the river and put it back where it needs to go. That's the idea. It is a positive, a break-even. It is not the answer but it sheds light on quantifying the process. This idea of overland flow is maybe a myth, an assumption that this was the case. After Katrina there was some accretion from sand deposits, at the same time as Katrina took sand away; a theory concerning Louisiana's lakes is that the bottom is in equilibrium with the fetch, and the wave energy stirs up the bottom and the silt ends up in marsh."

Obviously we were getting a little technical over lunch, but good scientists often relax together by bandying evidence about, thinking aloud and refining ideas. The situation was clearly sobering, regardless of the exact percentages. At stake were the lives of real people, the fate of New Orleans, and the future allocation of billions of dollars based on John Lopez's data points and the research of others like him (along with maybe some who disagree with his findings).

"Nobody says we can keep everything we have," he continues, "but can we preserve and build enough to keep the coast functional and protect our communities? Can we slow down the process soon enough to rebuild our ecological buffer and keep the coast from lapping at our levees?"

Like most good scientists, Lopez answers his own good question with, "We don't know. Sediment diversion, from an engineering point of view—can we construct systems to deliver sediment from river to marsh? If we build or rebuild a natural delivery system, we can see how it might work, but at this point we don't know. With sea level rise and subsidence, we can't know with certainty how it will all turn out."

To a hydrologist or an ecologist this was mighty pretty dirt. Yet wouldn't subsidence eventually send the rebuilt ground underwater as well? "This is an area of what might be called 'thick ice,'" answered Lopez. "It has resiliency because of the unusual natural river connection. The Gulf Coast without the river connection is essentially vulnerable—thin ice."

At the end of 2016, we revisited Lopez and the breach with a phone call. "Little islands are building up," he said. "The river water at Mardi Gras Pass is putting sediment down in a depositional process. It's clear what's happening. Basically, the water being captured in this outlet is carrying various amounts of sediment. Some of that represents erosion in Mardi Gras Pass, but most sediment is from the river. There's a canal at a 90-degree angle. It splits and some goes north and some south, and also secondary canals, but a lot of the sediment is due to the back levee canal. We measured the turbidity in the water. The deposition on the bottom fits the classic model of a discharge

plume in sediment, not because the water is influenced by various canals, but [because] it covers a large area and then five miles out it's
gone. Because of the back levee the canal distributes water and is
more likely to build land."

Sometime later, Steve asked Jeremy what he made of Lopez's
measurements among the spiders and the good earth. "I think we
were seeing how this natural levee was helping to stabilize and build
up slightly a little patch of ground. But everybody seemed to have a
different opinion, and when you don't know yourself, to take sides becomes an emotional thing rather than an evidenced-based conclusion.
But since there is no more silt coming down river, it's all moot. People
grasp at straws. It is their way of life. It's their home. They are ready
to be sucked into any perspective that tells them it is not hopeless,
whether true or not."

"Lopez?"

"No, Lopez is a real scientist, and he has heart."

Game Night

Short of a major hurricane, few events in southern Louisiana are
as important as a football game between longtime rivals Louisiana
State University (LSU) and the University of Alabama.[16] Appropriately for such a holy event, vestments are worn, bright purple and
gold for the mighty LSU Tigers, blood red for the Crimson Tide of
rival Alabama. Stakes are always high, especially when it's for the
national championship, as it was at the end of the 2011 season.

Spirits Hall was a sea of gold when we walked in after parking the
boat, stuffed with a raucous mix of oil and gas workers; oyster fishers; fly-fishers from Washington, D.C.; orthodontists from Jackson,
Mississippi; lawyers and private wealth managers from New Orleans;
professional conservationists; and even some members of a French
film crew filming a documentary on hurricanes. Several of Foster
Creppel's kitchen help were clearly cheering for Alabama (African
Americans at Woodland tended to root for 'Bama), and he told them

in front of the others that they were quite welcome to wear red. This was only a game, remember, only a game.

Jeremy, Steve, and Gene Turner were not great fans of football (they did not admit this outright). Jeremy, a New Yorker, was more of a baseball fan. Steve, from Montana, knew bird hunting better than football. And Turner, a Buddhist, kept his usual scientific calm. But even they got caught up in the excitement. Everyone was in high spirits as the big game began. But in the end, the Crimson Tide rose like a mighty tsunami and swept the Tigers away.

There was no crying after Spirits Hall emptied out. About the only sound heard was ice cubes crunching in Lucky Frank's mouth as Foster tossed them to him over the bar, one by one. "Can't win 'em all," said Foster, philosophically. "And how was your guys' day?" he asked, as he poured himself a stiff French Connection—Grand Marnier and Hennessey. "By the way, I got y'all set up for tomorrow—oysters and shrimp."

Oysters

A bustling mix of Cambodians, Croatians, African Americans, and Cajuns are unloading and sorting oysters along the open cement deck at Empire, three miles south of Woodland, when we arrive the next day. Tiny open boats pull up to the piers, full above the gunwales with burlap sacks that bulge with oysters raked off earlier in the day with hand hoes, and bigger boats sidle up with their holds full of industrial drags.

A tiny woman in an orange baseball jacket, pink pants, and tennis shoes, with a white and blue scarf wrapped around her face (all but the eyes), sits on the gunwale as her husband, in jeans and an LSU sweatshirt, waits to unload. They are Cambodians. Like many of the last-to-arrive oyster farmers, they work the unleased public reefs.

They sell to John Tesvich and his brother, who own AmeriPure oysters. The brothers are among the biggest oyster farmers in the region. Their father came from Croatia in the 1930s and worked from isolated camps in the bayou until he married their mother. Tesvich, a tall, fit man who just turned sixty, had studied to be an engineer but decided he liked the rougher life on the open bayou. It has made him wealthy, given him some reservations of late, but never slowed him down: "You try to get out on your bed at 4:30 in the morning, work it till maybe 3 or 4, come in, unload, watch a little TV, and go to bed."[17]

We step onto his boat. Sixty feet long, steel and aluminum with sleeping quarters, it is not luxurious but efficient, like a long-haul-trucker's cab. Tesvich's hobby is marine architecture, and he and his brother design their own craft. The boats cost $250,000 to $350,000, as much as a large tractor in Iowa.[18] He talks about how oysters are grown in Croatia on ropes, a very different kind of technique from that used for the natural and seeded beds on the bottom of Louisiana bays.

Half of Louisiana's $41 million annual oyster crop comes from Empire and nearby docks, most from Plaquemines Parish, second from St. Bernard.[19] For his AmeriPure brand, Tesvich pioneered a method of pasteurization to kill *Vibrio vulnificus* and *Vibrio parahaemolyticus*, the bacteria that can give upset stomachs to those who slurp oysters on the half-shell. The freshest and largest oysters command the highest prices. State law requires that the temperature of each batch be taken hourly on the boat. This is "virtually impossible" to follow on the way back to the docks, so Tesvich's boats are equipped with expensive coolers. These big boats—Tesvich owns another sixty-five footer—use winches and toothed baskets called rakes to "tickle the oysters off their bed. You don't want the rake too heavy and you don't want to pull it at the wrong angle, or you will drag up mud and bottom and have to wash it off," he explains to us.

The beds are reefs of amalgamated oysters, living or empty shells with a bottom of cultch (usually fossilized shells from sand mining). Once these reefs ran parallel to the shore for several miles, as they also did in the nineteenth century off the coasts of New York and Baltimore. Oyster reefs serve as a natural first defense against hurricanes and storm surge by slowing the water and breaking the waves.

Another big oyster boat motors past Empire's dock and Tesvich shouts out, laughing, his educated engineer's accent suddenly submerged in Cajun patois: "Hey, whar you gonna go with mah oysters? Come back hyra!"

Tesvich recalls the devastation after Katrina. He, along with his wife and his brother, had taken their big boats a hundred miles west to St. Mary Parish, "out of harm's way," the day before the big blow. When they motored back to Empire, the canals were littered with sunken craft, the buildings were flattened, and two feet of water coursed over the cement.

"There was a dead body pulled up where you are standing," he says, "The sheriff came by, ashen faced, on an airboat, and he said, 'John, it's all gone.'"

On the dock today, workers load sacks into a cavernous eighteen-wheeler labeled Rick's Seafood of Cross City, Florida, driven by a cheerful Alabaman with a tattoo on his right bicep showing the head and antlers of a massive whitetail deer, a trophy worth commemorating in permanent body ink. The man expertly shucked a couple fresh ones for us using an impressive folding blade. Jeremy pronounces these among the best he's ever sampled, and has four more in rapid succession. Not well acquainted with the pleasures of oysters, Steve found them slimy but filling.

We met for coffee at Tesvich's house. John and his wife built it on pilings in accordance with government recommendations for a hurricane storm surge, so it rises fifteen feet off the pad. With irony, he points out that Katrina's water, nevertheless, broke down the door and crested at thirty-seven inches in his living room, making a "grass line exactly fifteen feet and thirty-seven inches off the pad." He runs his finger along the painted-over stain. "But our house survived," he shrugs. "It didn't get washed away, I think, because we are on Nairn Ridge with some land and oak trees, that the fetch and the wind were lower here. Most houses were gutted."

It's not the water from Katrina that rankles him today, however. It's the idea of engineered freshwater influx destroying the Plaquemines Parish oyster beds, and his industry with them. "They want to build these diversions at 250,000 cubic feet per second and flow the river into Barataria Bay. You know what that does to the salinity in an estuary that is only ten miles wide? It becomes a freshwater delta. No way oysters can live there."

"When I was teaching at Johns Hopkins, back in the seventies," Jeremy commented, "a hurricane smashed into Chesapeake Bay and basically flushed the toilet all the way out to the ocean with fresh water, killing all the oysters."

"Yeah, but that was a once every twenty years event," John points out. "This would be permanent. These diversion characters want to create a new Delta—environmentalists, fishermen, hunters, the State, and alligator farmers. Fresh water's good for them. Well, we have a Delta south of here. I want to rebuild that Delta down there and preserve our estuaries up here. Build up the barrier islands down there. Import landmasses. They want to create freshwater lakes. They want to clean up Lake Pontchartrain. Don't put that dirty water in our backyard. They want to clean their lake so they can go sailing on it, but they want to mess up my sound. I have harvested those reefs. I have worked ten hours a day with a cane pole. Oystermen pole to feel the bottom, to feel for the reef. You find the edge of that reef because the oysters are usually on that edge between the mud line, what we call the roll of the reef. And then you mark that edge and then you harvest that edge. You are feeling those reefs, and I have done that for so long that I feel a kinship to the reefs. I want to protect them. I don't need to work oysters anymore for my goddam livelihood! And I have told this at the coastal restoration meetings. It would be a damn shame if our state allows the destruction of natural reefs like that. Our state is not aware of it. People in the Environmental Defense Fund, they are not aware of it. I am trying to make them aware. Now we have side-scan sonar of that area and we see those reefs."

Tesvich is on a roll. "I say well what about protecting marine assets like they do on land? We have communities on shore, we have oil refineries, what about natural assets in the estuary? Our oyster reefs are like the coral reefs in Florida. Nobody should be fooling around with wiping them out! But here they are thinking about putting up a huge diversion and talking about making a new ship channel, too."

In the kitchen of his elevated house, Tesvich fumes in his calm, fast, careful, erudite way. "Everyone is talking about 'restoring the Delta with fresh silt and too much water,'" he continues, but they are avoiding "the big elephant in the room, which is our oil industry. There is nothing in the master plans for coastal restoration about stopping pipeline dredging or what to do with all the open dredge channels and the canals that we already have. No mention. No blame. It is so sanitized, it reeks!" He continues: "And the environmental groups are just allowing it to happen. They want to turn this back to nature. They want a six-hundred-year plan, let the river go and let it go back to nature. They don't care about what humans are doing here. Everybody's afraid to tackle the big dog, oil and gas. They just want to reroute the river and allow this river to flow so that Louisiana will still be here six hundred years from now. They are not worried about what is going to happen between now and a hundred years out."

Tesvich fumes, "These diversions will kill the oyster reefs. The geologists, they see a shallow bay, like Barataria Bay or Black Bay. They think, 'Why not create more land? Fill it up!' I tell them, 'You stupid fools! You want to waste this kind of productive estuary for land? That wasn't land for five hundred fricking years. It never was land. Those reefs have been there for a couple hundred years at least. And you want to fill that in and just silt it over?' It doesn't make sense." He pauses politely to offer us more coffee.

"Is there a way to build up the Delta and the land without affecting oyster farmers?" Steve asks.

"Yes," insists Tesvich, "there is a less destructive way. Put the sediment in pipes and put it where you want to. Fill up the canals. Pump in sediment where you want it. It is more costly upfront, but you get results a lot faster. And it can help us in our own time."

"So I know you are a really young guy," laughs Jeremy, who was seventy-one then and proud of the emeritus after his name, "but when you retire are you going to retire in Louisiana? Or are you going to go to Croatia?"

"I am one step out of here, yeah," Tesvich replies. "I'm glad my son is going to engineering school. I want him to graduate, not follow me. I would go to Croatia. I would go to California. Somewhere the hell out of here. I am not going to stop in Baton Rouge. This is why I live here. I can go down on the dock. I can pick up a bushel of crabs off one of the boats, and oysters and shrimp right off the dock. That's the great thing about living here. You ruin that—you ruin our estuaries, what the hell? then—no, I don't want to live here in Louisiana. It is not a good state otherwise. That's the good thing about this state. Maybe music and food, too," he trails off, considering the state's exquisite cuisine and vibrant musical culture.

Shrimp

"What I always just say, you know, we got a rabbit starting from the Gulf running towards us and we got a turtle going out to meet him, and you know where they're going to meet, huh? Just a few hundred feet off the levee!"[20] Erik Hansen, the shrimp broker in Port Sulphur, two miles south of Woodland and five miles north of the oyster docks in Empire, leans back in his chair and laughs. "I'm confident that Mother Nature's going to keep providing us with shrimp," he says, "but as these estuaries keep getting smaller and smaller, there's no cribs for the babies. There's no marsh for the juveniles to grow up in."

Shrimping was down some 40 percent in 2012, he says. That's the short-term picture. Hansen blames colder water, not the lingering effects of the BP spill—nor the shrinking marsh estuaries, which he does worry about, on the near horizon and long term.

"See any diseased shrimp?" prompts Gene Turner, who joined us on this trip. "I've heard reports of black eyes."

"Nope, in fact, they were a little bigger this year, fewer but bigger."

Hansen leans forward, elbows on desk. He's a trim, agile man with a mustache and a pleasant Norwegian-French face more red than tanned. Hansen's people go way back in Plaquemines Parish. Cubit's

Gap was named for a great-great-grandfather who jumped ship over a century ago, set up a hog farm, and dug a ditch to bring the pigs water. The ditch became a ragged canal, and the canal a pass-through to the open bay.

Hansen, who sits on the board of the Louisiana Shrimp Association, is wearing jeans; the ubiquitous New Orleans' Saints T-shirt, long-sleeved and gray; and a clean white ballcap. He talks to us eagerly. "I'm making the same money per pound as my dad did back in the 70s," he says, "but we're moving probably a tenfold difference in volume. You see, back in those days, you know, a typical boat was maybe twenty-eight to thirty feet, the bigger ones, pulling a little thirty-five-foot trawl with a little 110 HP Fairbanks diesel, and now I've got guys with twin 500 HP Caterpillars in 'em. They move their rig a little bit faster, cover more area, so I mean, economics is going to take care of the size of the fleet. Most of the American guys have just moved on to something else. They just can't feed their families doing it. The few guys that still do it have to do winter jobs. You see, back when I first started, back in the 80s, man, a guy that owned a double rigger had a good life. He shrimped in the summertime, a little bit in the fall and the early winter. Man, from the beginning of December till April he just worked on the nets, did the housework, and all that stuff. Now, as soon as they tie up their shrimp boat, they've got to get on a crab boat, or they've got to get on that little oyster boat and go work down at a refinery. There's no such thing as a full-time shrimper anymore. It's not that lucrative, you know."

"My father was an oyster guy," Hansen continued, "but he always kept a shrimp boat just for part time stuff, so I was lucky in the summertime, and I'd go make a couple hundred dollars a day sometimes on that boat and I always had money in my pocket. I had the motorcycles and the cars and stuff like that, but I worked for it. It wasn't given to me!"

He laughs, "I'll tell you what, my dad and my mom begged me to go to college, but I was, like, on the docks with my father since I was five years old, and I just had a taste for it, and once I started making a few dollars, I was not going to go to college a year or two like a lot of the boys did down here and waste their parents' money, and drink and, you know, have a good time. I don't have any regrets. I'm fortunate in a way. We own this property, everything's paid for, and so I'm in a position where I can sustain a lot worse seasons than a lot of these guys. There's not too many shrimp docks down here that own their own property, so to speak, [and] don't have to pay leases and rents and stuff."

Hansen's office is tiny, a shed on the docks with a cash register, but he is one of the biggest volume brokers in Louisiana. "We used to just go over to the bank on payday and get cash for twenty-four guys and hand it to 'em. Now we got a computer for all that," he says in his soft, ever-polite drawl.

As a broker, Hansen buys the catch from the boats that dock in Empire. He then brokers the tonnage to the canneries. Hansen's not vertically integrated like Tesvich and the AmeriPure oyster company. At the dock, wholesale, the shrimp business in Louisiana was worth some $180 million in 2014 (about $126 million when we first talked with Hansen in 2012), making it a far bigger industry than oysters, yet only a fraction of the size of oil and gas.[21] But shrimp, like oysters, are an iconic part of the state's "Who Dat Nation" fleur-de-lis image. More importantly, shrimp, like oysters, require an intact and healthy marsh ecosystem, while oil and gas "don't care."

Erik Hansen may be doing all right today, but he worries about the future. "We have a problem with subsidence. There's a lot of minerals, you know, a lot of oil production in the parish, so, naturally, everything is sinking. Before they had the levees, the fresh water would provide the nutrients and all to sustain things and, of course, we've leveed everything off. With the oil companies coming in and dredging the canals, and the loss from the hurricanes and all, there's not a lot of meat left, just points and peninsulas. That skeleton can only hold on for so long.

"Everybody has had a hand it, the oil companies, the levees, and the boat traffic is another thing, too! When the Freeport Canal was dredged, and it's eleven miles long out to Lake Grand Ecaille, it was only about eighty feet wide, and, oh man! It's like six hundred to seven hundred feet wide now."

To Hansen the point of lost marsh is lost shrimp: red, white, and brown, the three commercial species of the genus *Penaeus* that collectively are known as Gulf shrimp. The browns are of less concern to him because they aggregate far offshore. What's killing offshore shrimping—the kind of shrimping seen in the movie *Forrest Gump*—is simply the price of fuel, which can run up to half the cost of a trip for the bigger boats (although that might change with falling prices). The dead zone forces brown shrimp even farther from shore, so that it costs still more to go out and net them. These days Hansen estimates that two-thirds of both offshore and inshore fleets are operated by Vietnamese immigrants, with perhaps 20 percent of the Vietnamese being of Cambodian descent.

"The shrimp bury up in the sand because of the predators," explains Hansen. "That's why they come inshore with their eggs, because they have protection in shallow water. The little white shrimp, especially, when they come into the estuaries, they bury up. We go for periods of time where we see 'em, then they totally disappear for a couple of weeks. What they do is they seek the shallowest water they can and they go in the mud, and they can survive, you'd be surprised, for a pretty good long time."

Hansen shakes his head when he describes the three-hundred-foot dredge ships under contract to the Army Corps of Engineers that keep the Mississippi open and deep enough for navigation. "They take all that good sand and sediment and dump it out in the deep blue Gulf," says Hansen. "Why can't they put it back on shore? It would be cheap, cheaper than what is coming to be known as 'The $50 Billion Solution.'"

Yet Hansen is a big proponent of building up the barrier islands. "Whew! They moved some serious sand out there," he marvels, referring to the massive transfer of dredged sand onto shore and boundary lines. What he doesn't like is diverting river water in an attempt to layer upstream silt onto the marshes. He believes river water is too fresh, so diversions just don't work. They change the delicate balance of salt and freshwater, which young shrimp and oysters need.

"They seem to kill the marsh around the outflows," Hansen observes, then laughs again. "Yeah, matter of fact, me and your host there, Foster, differ a lot on opinions on what's going on with this erosion situation."

He pauses for effect and a twinkle. "I'm a pipeline sediment guy. I say move it with pipelines and dredges. Then you address the problem, how much fresh water we're going to need to sustain it. But the idea of just putting massive diversions in, with unchecked flow? You can get one of these boys that live down in Venice, one of the local guys, and have him take you down to the mouth of the river where it's exclusive[ly] just fresh water flowing over, and you'll see they're losing their marshes as fast as we are up here! So fresh water's not the answer because it doesn't have a lot of sediment left in it. I mean, I'm not naive enough to think that we don't need fresh water, but I strongly think that we've got to build our coastline back up, our barriers, pump in and terrace, then divert a certain amount of water . . . to sustain it. This massive diversion they're talking about in Myrtle Grove, it's going to devastate us. Oyster fishing, crab fishing, the shrimpers, we're going to hurt."

He checks each of us with his eyes, then continues: "As you know, river water has so much nitrogen and other pollutants, that's what's keeping our oyster areas closed. That's the thing Foster is not aware of. When they get these big diversions like they're talking about in Myrtle Grove, the pollution line is going to get pushed so far out into Barataria Bay it's going to put some of these guys out of business. I mean, there's good stuff in that river, but there's a lot of bad stuff in there, too. Put everybody out of business!"

"How y'all doing?" asks Foster Creppel as we saunter into Spirits Hall later for hors d'oeuvres. "You catch a load of bull from those oyster and shrimp guys? They don't get it. They're all my friends, but they don't get it! This Delta was built with fresh water, and that's the only way to save it. Take out these back levees and let that good river water cover it. Now we got three kinds of shrimp tonight," says Foster sweeping an arm over the impossibly long serving table, a thirty-foot shallow wooden box made with old blond cypress planks, almost squirming with fresh boiled shrimp. But Jeremy is allergic to shrimp, and asks for oysters. Too bad.

You stand along the table, pull the heads off the shrimp and toss the shells back into the box—"here let me show you how to do that," offers Amanda, expertly shelling a fat white shrimp. "And what sort of beverage would you gentlemen be liking this evening?"

"Vodka martini!" Jeremy enthuses.

"Iced tea," says Gene.

"I'll have an Albita brown. Thank you."

Field research.

"Frank, get down!" Foster shakes his head, "That dog . . ."

We left Spirits Hall a bit late that night.

"I guess what I wonder when I hear all this stuff," Jeremy says as we walk the short distance from Spirits Hall to our rooms on the second floor of the Big House, "is what is the grand hypothesis here? What is the goal of all these diversions that people have such strong opinions about?"

"Well," says Turner, "they believe that fresh water is good because it will build land, that nutrients will grow soil, or above-ground material, and not put it in reverse. The water people believe there is a significant amount of sediment coming down, but it's actually a small, trivial amount, and their rationale is 'it's so natural.' "

A half-moon glinted off the alligator ponds. An empty ship high in the water moved upriver, sounding its foghorn through the trees, as we climbed the stairs to our rooms.

The Grand Hypothesis Revealed

Lurking in all these great conversations with our wise and savvy innkeeper, the fishers, and the scientists were allusions to the mysterious $50 billion Master Plan to save New Orleans and the Delta, which was evolving as it wended its way through halls of government.

Nothing had been resolved because every special interest wanted something different from everyone else. The only consensus was that a magical $50 billion of federal money was needed to solve the problems that the state had failed to deal with until that moment.

Then, quite miraculously, on May 11, 2017, the Natural Resources Committee of the Louisiana State Senate approved a fifty-year updated Coastal Master Plan to protect and restore coastal Louisiana—its communities, culture, economy, and wetlands—all at the agreed $50 billion mark for federal funding. The plan allocates roughly equal amounts of money to engineered protection and ecological restoration. Eight hundred square miles of new wetlands are expected to counteract subsidence, sea level rise, and storm surge, and so reduce flooding. An annual savings of more than $8 billion is expected due to averted losses from coastal flood damage.[22]

The plan is largely the work of the Coastal Protection and Restoration Authority, which developed the first version in 2007. After the devastation of Hurricane Katrina in 2005, Baton Rouge created a single agency to develop strategies and coordinate work with federal, state, local, and private institutions to protect and restore coastal Louisiana. Over ten years it has constructed or maintained almost three hundred miles of levees and sixty miles of barrier beaches and islands, completed projects in twenty parishes, initiated 150 new programs, and "benefited over 36,000 acres of coastal habitat," raising $20 billion from a variety of sources to pay for these initiatives.[23]

Nevertheless, Governor Edwards wants to declare Louisiana's coastal erosion a state emergency. This is because, since 2000, twenty-eight hurricanes or tropical storms have hit the Louisiana coast. The *Deepwater Horizon* famously ruptured and burned in 2010, but more than sixteen additional large oil spills have occurred in Louisiana waters since 2000. Storms trigger leaks and spills. Katrina may have caused half those spills, and the ongoing *Taylor Energy Platform 23051* leak began during Ivan in 2004. Substantial funding for the plan comes from federal offshore oil revenues and the *Deepwater Horizon* settlement. Ironically, oil industry money, expected in 2017, made the plan feasible.[24]

Many approaches have been suggested, each with proponents and detractors. One emphasizes diversions that send sediment-rich water from the Mississippi and the Atchafalaya to rebuild soil in decaying wetlands. Breakwaters and floodgates are popular short-term fixes with shorefront communities. But although a $200 million floodgate would save $4 million in yearly flood damage for the first twenty years, it would cost $16 million for the next thirty years. Similarly, two breakwaters would build up about 640 acres of wetland over twenty years, but would still lose 1,500 acres over fifty years under worse-case scenarios—at a combined cost of more than $600 million over fifty years. We also need to remember that the Mississippi is always an active player in this game. In 2011, near Pointe à la Hache, it burst its banks and created its own "cost-free" sediment diversion. Only ecosystem remediation, such as marshland creation, appears to be relatively benign fifty years on.[25]

The plan calls for $25 billion for structural and nonstructural protection and the other $25 billion for various mechanisms of hoped-for restoration, with the lion's share of $22 billion for marsh creation and sediment diversion. It's all about tradeoffs. And politics. According to Edward Richards, a law professor at LSU, "If you look really hard at the master plan, it's really about levees. All this wetland restoration they're talking about is a fig leaf to buy off the national environmental community."[26]

But others look at it as a rare opportunity to build consensus. State Republicans and oil and gas executives bristle at the mention of climate change, but people watching the water rising are convinced. Jeff Carney, director of LSU's Coastal Sustainability Studio, said, "If someone were to come in and say, 'No, no, no, that's a liberal conspiracy,' they would say, 'No, it's not, there's actually no land there' . . . We're losing our land here for one reason or another, so it almost becomes not the most important debate."[27] The Coastal Master Plan, even with all its integral parts that still inspire debate, unites almost everyone in their desire to save the Delta. It is cumbersome and contradictory, but inclusive, and it has managed, at least at this point in time, to galvanize public will. That alone is an extraordinary achievement.

It's also the first time that the governor and state of Louisiana have publicly and officially acknowledged the magnitude of the impending catastrophe—and that climate change and sea level rise are real. Since nearly three-quarters of Louisiana voters believe they will be affected by loss of coastal lands, this is democracy at work. Yet at the same time, Louisiana's congressional delegation continues to oppose EPA regulation of greenhouse gases and fossil fuel production. They oppose the Clean Power Plan, which would significantly reduce emissions and enforce limits on carbon dioxide and other global warming pollutants. They also oppose tax credits for renewable electricity, and tax incentives for renewable energy. In other words, the congressional delegates don't get it that the political will is changing.[28]

Politics aside, and despite the fragile, hard-won consensus, the Coastal Master Plan as presently conceived faces an enormous financial shortfall. Fifty billion dollars is not nearly enough money for Louisiana, and all the coastal cities and towns from Boston to Texas will also be requesting federal support.

Finally, will it work? We don't think so. But before we explain we need to meet one more person, the social scientist Shirley Laska, whose big-picture overview may be the most realistic of all.

Forays

Having learned much about shrimp and oysters, and tasted more of both at the Plantation House, we would soon strike out on a series of forays to explore the Delta, traveling east along the Gulf Coast into Mississippi and Alabama, and spending more time in New Orleans itself. But first we ventured south along Highway 23 to the Gulf of Mexico.

Running down the middle claw of the Bird's Foot Delta, Highway 23 is a road pretty much stripped to the bone—even today, it is decorated with blown-out boats, scattered FEMA trailers, and a funky, raucous Who Dat Nation Roadhouse style ("New Orleans strippers on Tuesday nights!"). But it is not a route without sinew. At the highway's end, off Halliburton Road, rises the robust Targa Resources facility, a natural gas processing plant atop a scrubbed field of imported gravel and dead bushes. Underwater pipelines laid at the bottom of Bayou Mardi Gras lace the canals, which date from the early boom years of the 1920s and 1930s, the era of Huey Long and Leander Perez. The older canals widen out from fading spoil banks, which crumble into the brackish water. The newer canals are prim and trim with manicured borders. To the north, at Empire, the oyster boats are full with shell. Port Sulphur is where the shrimp boats dock—the American Sulphur Plant is long gone. A couple miles farther north, Navigation Headquarters, like Targa Resources, is built with its first floor jacked up on cement pilings high off the ground to allow hurricane winds to sweep through.

On the Bird's Foot Delta, pretty much the only facilities built to last are for navigation and oil—except for Woodland Plantation, where the Big House somehow has stood unshaken through a score of blows since 1832, and Fort Jackson, which was rocked hardest by Admiral David Farragut on his rush up the Mississippi in 1862 to capture New Orleans. (This action sealed the Confederacy's doom as much as did the Battle of Gettysburg.)

"Any place called Fort Jackson should be checked out!" laughs Jeremy, as we pull into the parking lot facing the Mississippi River, between Maw's Sandwich and Snack Shop and the Lighthouse Fellowship Church.

Standing in the center of the parade grounds at Fort Jackson, you have to imagine how it was before the oak trees. "Birds planted them suckers, maybe a hundred years ago," explains the docent, who sports dark shades and a retro Fu Manchu mustache above his Plaquemines Parish uniformed shirt. Before the birds inadvertently seeded the oaks, which then spread out over the thick brick walls that once protected twenty-two big swivel guns, the parade ground held the wooden barracks for the fort, housing some five hundred soldiers during the Civil War. Situated sixty miles from the mouth of the Mississippi, hunched on a tight bend above Bayou Mardi Gras with Fort

St. Philip across the river, Fort Jackson was viewed as the impregnable defender of New Orleans and the Confederacy.[29]

In those days, the docent told us, the river here was narrow. You could shout across the water to Buras. The fort was actually softly floating, since it was built on a lattice of bald-cypress boards over soggy ground. "We just float," he said, "You can only dig so much in the dirt here, then you hit water. That's why we got tombs. It's not like Texas, where you can dig six foot under and you stay buried. Down here we got tombs so you don't float up!"

But the soggy-bottomed fort was no laughing matter in 1862. As soon as Admiral Farragut's fleet conducted a somewhat experimental mortar attack from floating gunboats, water tanks burst inside the fort. Munitions were set on fire, clothing stores burned, and the Confederate soldiers retreated below ground to the muck and the water. Once the soldiers retreated, Farragut gunned his fleet past the two forts. The Confederates had put a chain across the river, but it had already broken under the weight of logs and other flotsam, and had been only partially repaired. Farragut's sappers easily broke through.

By all accounts it was a wild and brutal battle—guns, mortars, ramming ships, gunboats, and forts ablaze in the predawn sky—but within hours it was over. The men in Fort Jackson mutinied against their commander, who had refused to surrender. The South had lost twelve vessels, the North but one. Farragut steamed for New Orleans, demanding the city's surrender. Citizens burned the wharfs, blew up munitions, and shoved warehouses full of cotton into the river, but the city lay below the river even then. Farragut's ships had a commanding sightline across what was then the South's largest city (168,000 slaves and freemen), larger than the next four southern cities combined. All that remained was panic.

Though movies are not made of it, this is where the South lost the Civil War.

Sam Houston warned Confederate Texans just before he was deposed from office for his irritating wisdom: "Let me tell you what is coming. After the sacrifice of countless millions of treasure and hundreds of thousands of lives you may win Southern independence if God be not against you, but I doubt it. The North is determined to preserve this Union. They are not a fiery impulsive people as we are . . . but once they begin to move in a given direction, they move with the steady momentum of a mighty avalanche, and what I fear is that they will overwhelm the South with ignoble defeat."[30] We were reminded of this as we walked around the installation and wondered if, fifty years hence, Fort Jackson's gun portals would be portholes, and if by 2100 this threatened site on the National Historical Register would be completely submerged.

Just outside Fort Jackson, a large brass plaque memorialized an event as important to our own ecological investigations as the fall of the fort was to the Confederacy. The plaque, erected by the American Society of Civil Engineers, "founded in 1852," was dedicated to the memory of the Eads South Pass Navigation Works. This engineering achievement made navigation up the shallow, wandering Mississippi to New Orleans safer, with a passage deep enough and wide enough for very large bulk-cargo vessels. Captain James B. Eads walled up the sides of the South Pass with willow jetties and deepened its channel to thirty feet. Begun in 1875, the work took five years to complete, and Eads maintained the waterway until 1901. After the navigation works opened, the Port of New Orleans could handle bulk shipments to and from Midwestern agricultural and industrial centers. New Orleans became an important international port; the Mississippi became a commercial and industrial highway; St. Louis, Chicago, and Minneapolis thrived economically; and the riches of America's Heartland were distributed to the world.

So Eads prevailed, even against the Army Corps, who vociferously opposed his methods (the first, but not the last, time they got things wrong). But the constrained river now flowed more swiftly, carrying upstream sediment off the Delta and into the deep, and starving the marshes. Decades later, upstream dams built after the flood of 1927 altered patterns of silt deposition even more.[31]

An oceangoing fishing boat rusting in the borrow pit, left over from Katrina, marked the turnoff to Woodland Plantation—and we were back in our southern field quarters for the night.

Several days later we looped back to LUMCON, Port Fourchon, and toured the beaches of Grand Isle. Steve had walked these beaches two years earlier, when he had toured the Delta during the BP oil spill and flown out over the *Deepwater Horizon* in a small plane. Jeremy had not seen them, however. To an expert ecologist there was still much evidence of the spill—skulls of dead seabirds, for instance—but what angered Jeremy more was a strangely over-built highway bridge that arches like a steel rainbow far above the open water and canals. He called it the "Bridge to Nowhere." We stopped the car at the apex of its elevated approach. "Look at the patches of dead trees! The trees rotting out in the water, the cuts, the canals widening and widening as far as we can see." To Jeremy, it looked like what Turner had shown them from the air, only the different angle of perspective made it seem more grotesque.

"It's rather pretty to the untrained eye," Steve thought, if you as-sumed this was how things were supposed to look. But he knew they were not. To Jeremy, this was a tableau of ongoing devastation. We were looking at the remnants of a vast marsh as it was still bleeding.

"A wound that will never heal!" shouted Jeremy from the top of the bridge. "How much does a stupid bridge like this cost anyway?"

Our attitude was turning from flip to serious, as we gained insight and traveled the region. John Spain, the rather clever executive vice president of the Baton Rouge Area Foundation (BRAF), who was wearing white buck shoes, stunned us by saying that it was the foun-dation's belief that in fifty years, Baton Rouge, 120 miles north of New Orleans, would find itself on open water as Louisiana's southern-most port. There are few climate science deniers where big money is concerned, and the BRAF has grown to some $574 million in as-sets.[32] It was also clear from our meeting at the BRAF that rival Baton Rouge will not mourn too much if and when New Orleans slips into the sea.

Eventually, we headed back to Louis Armstrong New Orleans International Airport and went our very separate ways, to La Jolla, California, and Washington, D.C. We had started as outside observers, scientific carpetbaggers, and though we were learning much, we knew that no one could understand deeply the changes that were about to occur without being rooted in this soil for generations. Still, we told ourselves as we began to put things together, seeing things with fresh eyes and without a personal stake in the outcome can at times be useful for deciphering what is really going on.

New Orleans's Ninth Ward sits fifteen feet *below* the Mississippi River.[1] Everything else "flows" from that fact, and you cannot help but be reminded of it as you walk along the levee's banked berm and look down on the roofs of the homes. From that vantage, they look like toy houses in a sandbox. We had focused on ecology, geology, marsh, and dead zones, but now it was time to return to people, to Americans, and to those in New Orleans who endured the worst of Katrina, and likely will endure the worst of the hurricanes to come: the mostly African American citizens of the Ninth Ward.

A few of the houses in the Ninth Ward are new and fine and modern looking, funded and built with help from Brad Pitt, the generous actor with ties to Missouri who has a strong interest in architecture. Other homes are broken and abandoned even now, many still marked with a macabre, but practical, spray-painted yellow X beside the front door. After Katrina these crude off-kilter crucifixes alerted rescuers and cleanup crews as to who was inside and whether they were alive or dead. When you stroll by one of these houses, it is helpful to believe that nothing will go wrong again, to hold a politician's faith in the Army Corps of Engineers, or a stronger-than-perhaps-warranted belief in "The Dutch Solution."[2]

Whatever storm surge barriers may be put in place to hold out hurricanes larger than Isaac, one can't help but dwell on the basic fact that the river rushes by far above the head of any blithe Saints' football fan watching "The Game" in a lawn chair on the roof deck of a rebuilt house. We were having only a casual look on a calm, beautiful, slightly windy day, but the drama and terror of hurricanes past infused our walk with a ghostly feel, and perhaps some residual anger.

It's not that people weren't putting their lives back together, taking advantage of rebuilt or subsidized new homes. If you click on the real estate site Zillow.com, you can scroll through beautifully practical and platinum-certified LEED (Leadership in Energy and Environmental Design) houses priced at $150,000, lying cheek to jowl against $36,000 three-bedroom homes that need a lot of work, usually listed as "For Sale by Owner." There are more rugged listings, too: perhaps for the urban homesteader tired of winters up north, there are simple 6,300-square-foot lots with abandoned slabs still covered with glass and twisted metal, and a listing that reads "Under Foreclosure." We walked and drove the Lower Ninth in a rented car that day, but you can use Google to view disaster and redemption from a laptop anywhere on earth and come to your own conclusions.[3]

We stopped to chat with a cheerful group of Texas church girls who were cleaning the brush off a row of abandoned lots adjacent to the slope up to the levee. They wore big loopy sunglasses above green T-shirts and wielded rakes. If someone was too sick or too old or still had not returned from Houston or Baton Rouge (seven years after Katrina), he or she could lose their lot.[4] It seemed strange to us that a Houston church group was doing the maintenance work that locals could and should have been doing, and strange, too, that the city would foreclose on its own people. But perhaps not so strange at all.

Shirley Laska

The night before Hurricane Katrina hit New Orleans, the yacht club on the lakefront not too far from Shirley Laska's house had scheduled a party for her neighborhood association. Laska was then vice chancellor of the University of New Orleans and had founded the Center for Hazards Assessment, Response and Technology. She is an environmental sociologist: an expert in how social factors cause environmental problems, and how humans then react to those disasters usually by rebuilding or relocating. She studies who goes where, whether or not they return, and why.

On Friday, the last day of the workweek, two days before Katrina struck, Laska took a call in her office. It was from a meteorologist friend at the National Weather Center in Slidell, Louisiana.

"This will be the real deal," he told her.

She believed him. Early Saturday morning she drove inland to Hattiesburg, Mississippi, and eventually stayed with a colleague in Lafayette for six weeks.

"But the yacht club partiers did not leave on Saturday," she explains to us over a bowl of roast duck in red curry at the Thai restaurant near her house.[5]

"Didn't those people have cars and planes and the money to leave?"

"Yes, and they waited until the next morning. They partied till dawn and then left. The yacht club was totally destroyed. But they had insurance, they rebuilt it, and it looks great now."

The yacht clubbers did the New Orleans thing, the smart thing, the party thing, because they were covered by insurance, as they would have been had they lived in San Francisco or Boston. Not everybody suffered in Katrina. In fact, Laska points out, the evacuation of New Orleans was the most efficient evacuation of a large city in American history. Just not everybody could leave or chose to leave, and once the federal government entered the picture under the Bush administration, the chaos began.

Laska wrote *Catastrophe in the Making: The Engineering of Katrina and the Disasters of Tomorrow* with William Freudenburg, Robert Gramling, and Kai Erikson.[6] The book's simple, direct, and oft-ignored point is this: Nature does not strike humans. Humans first strike nature.

"What happened to New Orleans is not a story about the way natural forces sometimes hammer us," the book begins. "Rather, it is a story about the way humans can rearrange the contours of the land they settle in, doing so in ways that make it, and hence themselves, more vulnerable and exposed—inadvertent authors of their own distress." Laska and her coauthors elaborate: "What we so casually describe as a 'natural' disaster is really a collision between some cataclysmic force of nature—a wind, a tide, a quake, a conflagration—and a human habitat located in harm's way."[7]

New Orleans is located at the bottom reach of perhaps the world's most fertile stretch of farmland, the nexus through which the wealth of the Heartland is distributed to the world. The city was once one hundred twenty miles from the open Gulf, a hundred plus miles of uncut buffering marsh, cypress trees, and oyster reefs. This made it unusual. Manhattan, for instance, has always been on the Atlantic (though it, too, was once buffered by oyster beds and some marsh). Laska and her book collaborators hark back to historian Ari Kelman, who points out that even though New Orleans was more protected, it was still "the nation's most improbable metropolis," always located on "an almost unimaginably bad site"—in the path of hurricanes, from the one that destroyed the Spanish fleet in 1779, greatly aiding the British in Baton Rouge, to those we remember better, Katrina, Betsy, Gustav, Isaac, and the ladies and gentlemen to come.[8]

The Mississippi also used to flow at sea level through New Orleans before the addiction to levees—the most tragic consequence of which was the Great Mississippi Flood of 1927. This flood is still perhaps the costliest "natural" disaster in U.S. history—$135 billion for Katrina versus $160 billion for the 1927 flood, in today's dollars. That flood was recurring terrain for sometime New Orleans resident William Faulkner. It flowed as natural backdrop for *As I Lay Dying*, *The Wild Palms*, and "Delta Autumn," the final story in *Go Down, Moses:* "This land which man has deswamped and denuded and de-rivered in two generations so that white men can own planta-tions . . ." Decades later Faulkner would write that the "Old Man" paid "none of the dykes any heed at all" as he "gather[ed] water all the way from Montana to Pennsylvania . . . and roll[ed] it down the artificial gut of his victims' puny and baseless hoping."[9] We should all read more fiction.

Laska is well aware of how oil and gas canals weakened the marsh. "In the four decades between 1937 and 1977, approximately 6,300 ex-ploratory wells and over 21,000 development wells were drilled in the eight Louisiana coastal parishes," she writes. She explained how this engineering was done, with the barge-mounted draglines that Turner had described and clever submersible drilling rigs, too: small ships with rigs on board lowered on a likely spot and partially sunk. When the gas had been extracted, the drilling barge would be backed up through the marsh and routed to another likely spot. And there were a lot of likely spots.[10]

The 120 miles to the Gulf had been sliced and diced with a thousand cuts. But to Laska, the most unkind cut of all, even more than the endless oil and gas canals cited by Turner, had been "Mister Go," the Mississippi River Gulf Outlet, that seemingly robust straight blue canal that we saw stretching under us on our flight from Breton Sound in the Gulf westward almost to New Orleans. For Laska, Mister Go was the prime example of inadvertently engineering a city for its own eventual destruction. Mister Go seemed like a good idea at the time, that is, a good project promoted by the "Growth Machine," as Laska and her co-authors refer to the backers of heightened commerce and real estate development in New Orleans and most cities. Mister Go took so long to plan, develop, fund, and build, however, that it acquired a life of its own, only to arrive "obsolete on delivery," too shallow for modern oceangoing ships.

Mister Go was conceived in 1937 and funded in 1956, after Congress had passed competing appropriations for the St. Lawrence Seaway. Ground, or rather marsh, was broken in 1958, and the project finished in 1963. By this time, however, ships had become much larger, and Mister Go's usefulness was in doubt. Only twelve ships passed through the engineered canal in 2004. Each round trip cost the taxpayers of the United States approximately $1.5 million. In other words, Laska explained over a dessert of Thai fried bananas, Mister Go was a boondoggle. Some people got rich from Mister Go—mainly higher-ups in dredging and construction companies, as well as the canal's facilitators in Congress and the Louisiana legislature. Other people made decent livings from the project—for example those in the Army Corps of Engineers and construction workers. Others died.

Mister Go became the hurricane highway that allowed Hurricane Katrina's catastrophically heightened storm surge to funnel at great velocity into New Orleans through the hypodermic needle at the funnel's end, bursting the floodwalls protecting the Lower Ninth Ward and "shoving away the homes like a battalion of bulldozers."

To Laska, the die was already cast. The efficiency of the city's evacuation, the horror of the Bush administration's slow and inept response, the very strength of the hurricane itself, this was all after the fact, tragedy upon tragedy, since New Orleans in her view had already been set up to fail.[11]

"Floods are an act of God, but flood losses are largely an act of man," wrote Gilbert White, the dean of American geographers, decades ago. Though in the future, people may help God in regard to floods, levees today convey a sense of wellbeing that is not justified, especially when floods—or hurricanes—prove to be bigger than expected.[12]

The sins of Mister Go were twofold. First, it let the water into New Orleans, and at such a velocity that it overtopped the walls protecting the main part of the city and the Lower Ninth Ward. Second, digging Mister Go severely weakened the marsh protecting New Orleans to the southeast. It is a train of thought in keeping with Gene Turner's perspective. Bad engineer-think and politics trumped ecology and common sense. And none responsible said they were sorry.

Laska points out that 270 million cubic yards were moved to dig Mister Go, more dirt than was dug to excavate the Panama Canal. The channel could have been plotted with a draftsman's ruler. It was straight as a pin. During a major hurricane, water could rush straight up with a head at the front "like snow before a show shovel." In normal times, too, Mister Go became a slackwater ditch, allowing saltwater to enter the vast marsh area of the cut. Saltwater intrusion then weakened the roots of the marsh grasses at the edges of the canal, causing the banks to cave in, the channel to widen, and more saltwater to slide up on every tide—even more with each storm. Cypress trees were the first to go. As the channel filled in with new dirt from the sides, the Army Corps of Engineers needed to dredge it out. By the time of Katrina, the channel was half a mile wide—a true catastrophe in the making.[13]

Increasing salinity from the widening canal caused a regime change that helped brown shrimp, which were able to handle more salt, and hurt white shrimp, which were not. Fish species that preferred less salt disappeared. Oyster-growing moved inland, and oysters became more susceptible to bad bacteria. In the end, Mister Go would make shrimpers and oystermen leery of further channels and diversions that might hurt habitat, whether proven or not.

Mister Go, the blue ribbon that had appalled us from the air, destroyed an estimated 23,000 acres of marsh and raised the salinity to a grass- and cypress-killing level in 30,000 more acres. It also weakened approximately 120 miles of intact marsh that had once buffered New Orleans and St. Bernard Parish to the south from storm surges, including those from hurricanes.

In a triple whammy, the wetlands of the twenty-first century had become so weakened with oil and gas canals, shipping channels, and overly nitrated water from the Heartland that the city of New Orleans was left standing naked before the winds.

Cattails to Seagrass

Until now we had been gathering information, listening to different perspectives, and having a lot of fun. We had made friends. We had fallen in love with New Orleans and the Delta. But now it was time to step back and consider what we had learned.

The people we met are all smart, cool people who love their city and the region. They are invested in it, and they are serious about the goal of preserving its culture and surroundings. But they differ greatly about what they think are the problems and the solutions.

John Lopez, director of the Lake Pontchartrain Basin Foundation, has been studying the Bohemia Spillway. This was the break in the levee south of New Orleans dynamited by the city's bankers during the Great Flood of 1927. The incidental ecological result was to provide an eighty-five-year-old example of natural river flow. Lopez wanted to know if this overtopping would bring back a bit of solid ground. Would the deposited silt build up a natural levee as in times of yore? If so, this would be justification for the state to build huge pumps and siphons to suck mud and muck, sand and silt out of the bottom of the river and deposit it on land to build up the Delta further. More river water would add more silt, the ten thousand cuts of the oil and gas canals would start to fill in naturally, and New Orleans would have begun to rebuild its earthen barriers to storm surge and hurricane destruction.

Yet, and this has nothing to do with Lopez and his science, such projects would be extremely expensive and potentially vulnerable to graft. As someone said to us in a Delta bar, "We have a saying in Louisiana—if it's raining money, put out your buckets!" Whose money? The federal taxpayers, of course, not the oil companies that dug the canals, or the refineries that benefit so much from the current path of the Mississippi River. (An exception is BP, which got nailed for billions for its *Deepwater Horizon* spill.)

Foster Creppel, proprietor of Woodland Plantation, south of New Orleans in Pointe à la Hache—our southern field station—agrees with Lopez, and then some. Creppel's vision is practical, but steeped in romance. He wants to restore the Delta to the way it was when he hunted it, age nine, shotgun in hand, ducks tucked into his coat, and he is wise and thoughtful enough to believe that bringing back the river is what will do it. The slow turning of the snake's tail over thousands of years laid down the Delta before. It may take fifty, a hundred, or two hundred years he told us, but this is nature's way. So he, too, feels that the oil and gas businesses should fill in their canals, do the right thing, and when his high school friends, his fishing buddies in the oil and gas business, ask him, "Foster, why do you care?" he fires back, "Why don't Y'ALL care?"

Creppel is a proponent of the giant siphons that suck sand and mud from the center of the river and deposit it through slurry pipes onto islands and banks. He took us out in his boat to show us how this would build land. Along the canals he pointed out the middens and the remains of old Native American encampments. "They knew what high ground was! When all these spoil banks were put in, that changed the flow! That killed the trees! See those trees now by that midden? They're all dead!"[14]

Creppel's solution includes taking out the back levees. Let the hurricanes rip. Let the brown flood wash over the land. Good dirt will be deposited in alluvial minifans, just as it once was. The waters will recede naturally, and there will be ducks in the pond.

Just as it once was. Just as it once was.

Gene Turner has spent a lifetime in science pounding a stake into the atavistic vampire heart of that theory, though as we know, he remains good friends with Foster Creppel. Turner and Nancy Rabalais were the ones who proved that the water itself is no longer what it once was. Since 1970, Mississippi water has become full of nitrogen and phosphorus from Iowa and the other farm states upriver. The fertilizers and inputs that contaminate the river are mostly manufactured just upstream of New Orleans, in the giant petrochemical plants that constitute "Cancer Alley."

Remember Gene Turner jumping to prove that organic marsh is not solid muck but a tangle of roots that holds it all together? And how it's not sediment or silt? Dumping river water on it cannot bring it back. This is not just because Mississippi River water now holds only a fraction of the silt it contained before FDR, when the big dams were built in the Heartland and the levees-only policy of water control shunted river water far out into the Gulf, but also because the heavily nitrogenized water kills the roots of marsh plants, the matrix that holds the organic matter together.

So there is no point to the pumps, siphons, and diversions. For Turner, the crown jewel of wrong-headed thinking is Lake Poo Poo, where fertilized water killed the marsh instead of saving it, as we saw in our overflight. That overflight also crystalized our image of the profound sickliness of the Delta's marshlands, which have been lacerated by the open sores of oil and gas canals, old and new. These canals expanded from fissures to ponds, and ponds to lakes, and the salty Gulf intruding, killing more marsh, widening more grass wounds until New Orleans became desperately vulnerable.

This deadly irony, which so fascinates us, forms the toxic loop that is central to this book. Nitrogenized river water may increase plankton blooms that feed fish, but it weakens the organic marsh and creates dead zones in the Gulf the size of Rhode Island and Connecticut combined. It also causes toxic algal blooms that increasingly taint lakes in the Upper Midwest. As you may remember from Chapter 4, Toledo's drinking water was poisoned by a bloom in Lake Erie.

All of which brings us back to the Coastal Master Plan that, in spite of its hard-won consensus, we find to be deeply flawed. That's because any hope of success depends on two critical assumptions. The first is that there is enough sediment in the river to rebuild the coast fast enough to keep up with a rapidly increasing sea level rise. But there is virtually no evidence to suggest this is true. To the contrary, geology tells us that too little sediment in its mainstem caused the Delta to fall apart. We found no published scientific articles confirming that diversions of river water have resulted in a significant increase in land area. John Lopez's decade-long study of the Bohemia Spillway has demonstrated at best a few inches of accumulation. At worst, a decade of observations by the U.S. Geological Survey has demonstrated considerable land loss due to freshwater incursions through natural crevasses near Fort St. Philip. All the existing evidence suggests that the supply of inorganic sediments in Mississippi River water is insufficient to keep up with subsidence and sea level rise, let alone enough to rebuild land.[15]

The second assumption is that freshwater river diversions can restore normal marsh vegetation. Gene Turner has forcefully argued that this is impossible because of the river's excessive nutrient loading. A detailed analysis over nineteen years of three freshwater diversions onto marshland showed no significant changes in relative vegetation or overall marsh area from 1984 to 2005.[16] But after hurricanes Katrina and Rita struck, marsh area subjected to the freshwater diversions suffered dramatic vegetation losses. In contrast, losses on marshland not doused in river water were moderate and short-lived. Freshwater diversions are bad for marshlands, because, as Turner has shown, nutrient enrichment compromises the root systems that hold the marsh together. Turner is widely cited (he has more than ten thousand citations on Google Scholar). It should be profoundly disturbing to people living on the coast that his work is not cited anywhere in the Coastal Master Plan. This lack of scientific objectivity and common sense is counterproductive and undermines any hope of success.

Indeed, cynicism suggests that the Coastal Master Plan is about Band-Aids paid for by the federal government that may help in the short term, but become fatal in the long term. The greatest flaw in the Coastal Master Plan is a lack of skepticism—what if it fails? What is Coastal Master Plan B? Under reasonable assumptions of sea level rise, almost everyone will need to move north, sooner than anyone admits today.

Shirley Laska's simple dictum needs repeating: Nature does not strike humans. Humans first strike nature. Besides the ever-widening canals, she draws special attention to Mister Go, the shipping channel conceived so long ago that, after it was finally dug by the Army Corps of Engineers, few modern ships could use it. It was supposed to make the levees of the Roosevelt era look as smart as the interstate highway system. But Mister Go was so dead on arrival that others were killed when the first big blow—Katrina—pushed the storm surge through the system, crashing it into the walls of the city.

"Floods are 'acts of God,' but flood losses are largely acts of man," as geographer Gilbert F. White famously put it. Laska's solution involves a lot of retooling, and she cautiously introduces that word dreaded by all who love New Orleans and many other coastal cities here and abroad—*relocation*—because the sea level rise coming this century makes all these problems worse. Much worse.

After we had returned to Washington, D.C., and to La Jolla to begin a draft of this book, Laska sent us the first of two thoughtful emails: "I would like to explain my answer about the future of New Orleans," she wrote, at first. "It is simple: Climate change will increase the sea level, [and] may exacerbate the force of hurricane storm surges in the Gulf at the same time as [the Mississippi Delta] continues to be lost due to the erosion caused by subsidence and saltwater intrusion. I fear that the three dynamics will challenge the one in one-hundred-year storm safety level of our levees beyond their capacity to provide protection. If we have another inundation throughout the city as we did with Katrina, investment in the city will dwindle, the population will diminish, large hurricanes will no longer be explainable as a single freak event and it will become a regional city primarily supporting the functioning of the port. It is not 'farfetched' to believe this scenario could occur. We should be fighting like crazy to reduce the continuing major risk, but instead we are now complacent in thinking we are safe. [New Orleans is n]o different from any other area struck by a catastrophe, at least in our culture."[17]

But a few hours later she wrote us a second email: "Hi Steve and Jeremy, My comments to you regarding the future of New Orleans [have] prompted me to think more about the ideas I proposed. I've decided that giving you a 'sensational' quote rather than a measured analysis will reduce the chances of locals paying attention if you were to use it in your book. It would be seen as too extreme and thus be dismissed. Instead I would like to have you simply say that, as with any place that relies heavily on levees, the levees act as a 'calming' action that diminishes community awareness of its continuing risk . . . Now that the city is doing so well in business growth, its I.T. focus, young people coming in, high growth in the local movie industry, etc., attention has definitely turned away from the issue of what continuing commitment must be made to safeguard the area from flooding. This is inevitable culturally, perhaps even humanly. I came to New Orleans in 1967, two years after hurricane Betsy, and hardly a word was spoken about Betsy. The better New Orleans becomes, the more we have to risk by such complacency."[18]

Shirley Laska nailed it. The facts are simple. They cannot be dodged or massaged or romanticized. The Delta is sinking and the seas are rising. You can slow the rate of erosion by filling in the canals, if oil companies and the state come clean and somebody pays for it, but that won't change the rate of sea level rise. Nothing will short of wholesale abandoning of fossil fuels for renewable energy. A "$50 billion wall" can be built around the city. But it will fail, even at that price. And federal taxpayers will want to save other cities that they live in, from Galveston to Miami to New York, perhaps by building more walls, which will balloon the $50 billion price tag (which no one believes is enough) into the realm of hundreds of billions to trillions of dollars. Indeed, as this book goes to press, the grim reality of all this after the one-two-three punch of Hurricanes Harvey, Irma, and Maria is finally sinking in.

The Mississippi's water will descend in pulsing torrents as the weather becomes more episodic in the Heartland upstream. Surge and rise will be unstoppable. Consequently, New Orleans may come to look quite different over the next decades, becoming a little buttressed island isolated from the mainland, peopled by ten thousand engineers and techies, flown in and out to manage ship traffic and petroleum processing—but outside that massive wall will be swamp all the way to Baton Rouge, and beyond that, seagrass and open water. The economic value of the port could be maintained like the facilities at the end of the Birds Foot Delta are today. Such a transformation would not be economically motivated; it would be a matter of human safety and wellbeing. In the future, the Louisiana exodus might become a symbol of courage and an example of how a vibrant culture and people survived.

PART III

Too Little Water and Too Much

The sea level is rising because the earth is getting warmer as carbon dioxide accumulates in the atmosphere from burning oil, gas, and coal. Warmer temperatures cause water to expand and ice to melt. None of this should be controversial. It's basic high-school physics. So far, most of the world's rise in sea level has been due to thermal expansion, but as warming increases the melting, the remaining ice on the continents is becoming more and more important. This was predicted with remarkable clarity in 1896 by Swedish chemist Svante Arrhenius, who called it the "hothouse" theory of the atmosphere. Arrhenius won the Nobel Prize in Chemistry in 1903.[1]

The earth's average sea level rose about seven inches between 1900 and 2000, and at least half of that can be attributed to increasing emissions of greenhouse gases. But now rates are accelerating even faster than the climate models predicted because warming has set off chain reactions that intensify the impact of rising temperatures. For example, in the Arctic Ocean more and more sea ice melts each summer. Sea ice covered with snow reflects back most of the sun's heat. But the darker seawater absorbs more heat, which causes it to warm even faster. As more sea ice melts, even more sea surface warming occurs, and as the melting accelerates, the rise in sea level accelerates. In the Arctic, vast stretches of frozen land are also thawing as temperatures warm. As this permafrost thaws, its soils release enormous amounts of methane, which is roughly thirty times more potent than carbon dioxide as a greenhouse gas. And that's not all. Methane also bubbles up from Arctic seafloor sediments. The release of methane accelerates warming even more, which causes permafrost to thaw faster and faster. And so on. You get the picture—it looks like a train wreck.[2]

These runaway processes are examples of positive feedback. And the more climate scientists look, the more examples of positive feedback they find. Feedbacks are difficult to model because of their many complex and accelerating interactions, but as scientists figure out how to do it, the fit of the climate models to what is actually happening is getting better and better. Far from being alarmists, scientists on the Intergovernmental Panel on Climate Change (IPCC) have been overly conservative in estimating rates of warming and sea level rise. That's of grave concern, because nearly half of the global population and three-quarters of large cities are situated within forty miles of the coast, and in Asia most of these large cities are barely above sea level. In 2010, 40 percent of the U.S. population lived in counties directly on the coast, and that's expected to grow to 50 percent by 2020. Roughly 2 percent of the global population lives three feet or less above sea level, including more than ten million people in the Eastern United States.[3]

According to the latest IPCC prediction, the sea level should rise between 29 to 39 inches by 2100 if emissions continue unabated and global temperatures rise by 5.8 to 9.7 degrees Fahrenheit. But the probability is only two in three that the actual rise in sea level will fall between those limits.[4] Given that no one seriously believes the change will be less, this means that there's a one in three chance that the sea level will rise by more than three feet under a high-emissions scenario. A one in three risk of a very bad outcome is cause for great concern. A good poker player would probably fold with a hand like that. A one in a hundred thousand risk that a car might hit children walking to school is unacceptable. We would install stoplights or post guardians at crosswalks to control traffic. But that kind of common sense hasn't been applied to the risks of sea level rise, or other scary consequences of climate change that until now have been easy to ignore.

Why is the scientific probability so low? The problem is that no one knows for certain how global warming is affecting the massive ice sheets of Greenland and Antarctica, which contain about 69 percent of all the freshwater on the planet. Because of this uncertainty, the IPCC has barely begun to include in its models how much the sea level might rise if and when the ice caps melt. Meanwhile, the news from the field is increasingly alarming. Meltwater is pouring faster and faster from the top and the bottom of the mile-thick ice sheet covering Greenland, which contains enough water to raise global sea level by 23 feet. Few scientists expect this to happen before 2100, but meltwater from Greenland already accounts for about 10 percent of the current rise in sea level. More alarming in the near term, new evidence shows that the Antarctic ice sheet is much less stable than previously believed. Antarctica is the behemoth of sea level rise, with just over two hundred feet locked up in its ice, so that anything that seriously destabilizes Antarctica is very bad news. Moreover, the new evidence suggests that Antarctica, on its own, could contribute three to four feet to the rise in sea level by 2100—more than the entire IPCC projection.[5]

Many different lines of evidence need to be considered together in order to understand what's going on. So Jeremy contacted his old friend Rob Dunbar at Stanford to explain. A Scripps-educated geologist and oceanographer, Dunbar studies rapid climate change in Antarctica and the Southern Ocean and constructs high-resolution records of climate and ocean variability. A veteran of more than thirty Antarctic expeditions totaling about five years on the continent, he received the 2016 Medal for Excellence in Antarctic Research from the Scientific Committee on Antarctic Research.

Dunbar explained three major factors that contribute to Antarctic ice sheet instability. The first is that the ice sheet is not directly grounded everywhere on the underlying basement rock. Instead, it sits on a slippery, highly dynamic interface of four hundred lakes that lie between the overlying ice sheet and the ground below. The largest is Lake Vostok, which sits more than thirteen thousand feet beneath Russia's Vostok station, located on the East Antarctic ice shelf. About the size of Lake Erie and with an average depth of 1,400 feet, Lake Vostok is visible by radar from the air. Penetrated by drilling some years after its discovery, the lake is filled with liquid water. The lakes are connected by a network of subglacial rivers, and can empty or fill in a couple of months, especially where the ice sheet is grounded below sea level.[6]

The discovery of liquid water was unexpected, but we now know it results from pressure melting, and the fact that the bedrock under Antarctica is a bit warmer than was thought due to heat from radioactive decay in the core of the earth and to volcanism. Amazingly, there are active volcanoes on Antarctica deep beneath the ice.[7]

Dunbar gives us some background on these volcanoes. "Mount Erebus would be the best-known example. but it's not the only one. One of the places I've taken people on field trips over the years is Brown Bluff on the tip of the [Antarctic] Peninsula. It's the result of a tuya, a supersized eruption under a thick ice sheet. And so enormous amounts of heat, enormous amounts of ice, and you get a lot of meltwater. The meltwater forms fast. There's a lot of pressure and it just starts squirting out from under the ice sheet. And as it does, it takes all of the products of the volcano erupting under the ice, all this ash, huge boulders, little grains, all this stuff gets moved, drawn together by these crazy currents under the ice sheet, and then it starts making these big conglomerate beds. You know, the cliffs at Brown Bluff are 1,700 feet high."[8] Brown Bluff is not the largest volcano on Antarctica. It formed within only the last one million years and several other parts of Antarctica also have the potential for modern volcanism.

The clear lesson from all this is that the ice isn't anchored solidly to the bedrock.

The second take-away is that an enormous amount of meltwater forms and flows on the glacial surface until it descends through cracks, crevices, and gigantic, roughly circular moulins into the depths of the glacier. Most of what we know about meltwater comes from Greenland, where everything is more accessible and scientists can routinely land on the surface by helicopter.[9]

"I was up on the ice cap in Greenland last August on a hot summer day. Although it felt a lot cooler it was well above freezing, in the mid to upper 40s, and there was ice water everywhere, big pools . . ."

Jeremy interrupted: "Can you hear it?"

"The roar is associated with the moulins, you know, so, if you're really close to a big moulin it's like a jet engine because [the meltwater is] just falling into the abyss. We landed by helicopter in a place the pilot said it should be safe, but you could hear the sound of water moving through fractures."

Wherever the water descends from the surface, the cracks and crevices become bigger. This, in turn, causes "hydrofracturing" of the glacier. Until recently these phenomena had only been documented from Greenland. But then, in 2017, Jonathan Kingslake and colleagues demonstrated that the same thing is happening all around Antarctica. They painstakingly compiled visual aerial and satellite imagery obtained between 1947 and 2015. Surface drainage has persisted throughout the entire time series and may extend seaward as much as seventy miles across ice shelves from the landward border of grounded ice. This abundant meltwater is a major factor in the breakup of large areas of the West Antarctic ice shelf.[10]

The third factor relates to what is termed the "marine ice sheet instability hypothesis." Ice shelves float on seawater, so their melting contributes nothing to sea level rise. But when they break away from the mainland, they set off chain reactions with dire implications. Consider West Antarctica, where a new iceberg the size of Delaware broke away in July 2017. The ice shelf is only 250 feet thick at the margin, and most of that's underwater. But it thickens toward the coast to where it's grounded on the bedrock, often well below sea level—in some places five thousand feet down and as much as a hundred miles inland.[11]

This grounding line is where hydrofracturing works. Rob explains: "So the ice comes down off the continent, it gets to a point where the base of the flowing ice lifts up off the sediment bed [into the ocean], and it leaves a grounding line of sediments. From that point on, it's flooded and hit by tides. Tides are working it up and down. It's mostly isothermal because it's sitting there on ocean water that can't be below 28.4 degrees Fahrenheit. And the upper part can be −20 to −40 degrees in the wintertime, but still heat's moving up through the ice shelf and the tides are moving it up and down, so . . .

it experience[s] brittle fracturing just from what the tides are doing. Then you get a strong surface meltwater event, and the rivers that are coming down off of the ice sheet . . . flow down on the ice shelf and hit those tidally induced fractures and the next thing you know it's all falling apart. That is how the Larson B Ice Shelf degraded in a period of 30 days in 2003. I've seen the photos, an Argentine friend of mine was down there then, they figured out what was going on, we got some aircraft down there and we got some overflights, and it's just insane. The whole shelf, just gone in thirty days." This breakup is rapidly continuing; another iceberg the size of Delaware broke away in July 2017. The fracture that gave way was more than 120 miles long.[12]

The places where the ice shelves are grounded to the bedrock act as buttresses that impede the seaward flow of the glaciers that lie behind them. But the bedrock slopes downward away from the coast because of the incredible weight of the ice sheet above. So as the ice shelf melts backward, it's not grounded anymore. This results in rapid melting and breaking up of the shelf as it retreats to a new grounding point, which may be fifty miles inland from the previous coastline. The loss of the buttressing effect also allows glaciers to resume flowing seaward, which causes additional sea level rise. In those cases where the new grounding point is way below sea level, a previously submerged vertical wall of ice is exposed to the warming Southern Ocean, which further contributes to a rise in sea level. And on top of all that, every bit of new sea level rise tends to lift the margins of the ice sheets off the bedrock, which accelerates all of these processes even more—another example of positive feedback that increases the dangers of climate change.[13]

The science is advancing incredibly fast, but our observations span only a few decades. That leaves skeptics room to foster doubt. But historical records from two key studies show that it has all happened before. The first stems from the ANDRILL project that cored the West Antarctic ice sheet near the U.S. station at McMurdo. They found that the ice sheet had advanced and retreated at least thirty-four times in the past five million years due to fluctuations in temperatures related to variations in the path of the earth's orbit around the sun, called Milankovitch Cycles. The second was the 2010 ice-coring expedition of the International Ocean Drilling Program (IODP) aboard the *JOIDES Resolution*, designed "to explore the theory that the single most important driver [of the initial formation of the ice sheets] was the decline in atmospheric carbon dioxide, rather than other factors such as the opening of circumpolar seaways." The team drilled at a site immediately offshore of a glacial trough five thousand to six thousand feet deep that extends all the way inland to the South Pole.[14]

Dunbar concluded: "Some of these retreats occurred when the rise in temperature was less than 4 degrees Fahrenheit. Almost everyone thinks we've got that much in the pipeline already. Think about the time scale of removing the excess CO_2 from the atmosphere and running through the various biological and geological cycles to remove it. There's a lot of inertia built in. So if we have a way to geo-engineer CO_2 out of the atmosphere fast or if we can mess around with deflecting some of the sun's rays with reflective particles in the stratosphere, okay, maybe, but without that it could be worse. That's the reason to go whole hog on these things."

Any confidence that the sea level rise will be less than three to four feet by 2100 is diminishing fast. Independent analyses of the work of thirteen ice sheet experts, incorporating observations from Greenland and Antarctica, showed a 5 percent probability of almost a six-foot rise in sea level rise by 2100 under the IPCC emissions scenario. Two additional independent studies incorporating polar ice sheet data reached similar conclusions. With the new insights about Antarctica, the probability of a six-foot rise by 2100 increases even more. What this means is that most of Miami and New Orleans—not to mention low-lying coastal areas from Maine to Texas—will be entirely underwater in about fifty years. In the jargon of statistics, this kind of probability distribution, which shows the likelihood of an extreme event, is called "fat tailed." Fat-tailed distributions appear to be characteristic features of the climate system and have very serious public safety and economic implications.[15]

What does five to six feet of sea level rise by 2100 mean for familiar places? The *New York Times* published a set of interactive maps in 2012 showing what parts of twenty-five major coastal U.S. cities would disappear with five feet, twelve feet, and twenty-five feet of sea level rise.[16] The maps were based on elevations from the U.S. Geological Survey (USGS) and tidal data from the National Oceanic and Atmospheric Administration (NOAA). These much-maligned agencies have been busily compiling reliable data and developing knowledge critical to any well-informed response to climate change. The data have always been publicly available on the agencies' websites for use and comment, though these are websites the Trump administration is now shutting down.

Just a five-foot increase is truly alarming. The worst hit would be Miami Beach and Miami (94 percent and 20 percent flooded, respectively), New Orleans (88 percent flooded), and Atlantic City (62 percent flooded). After these, St. Petersburg and Tampa (32 percent and 18 percent), Cambridge and Boston (26 percent and 9 percent), Virginia Beach, Norfolk, and Newport News (21, 9, and 8 percent), Jersey City (20 percent), Charleston, South Carolina (19 percent), Wilmington, Delaware (11 percent), Tacoma, Washington (10 percent), Savannah, Georgia (8 percent), and New York City (7 percent, including all of LaGuardia and JFK airports).[17] In general, Florida, Louisiana, and New Jersey will inevitably suffer extreme losses of land, but all along the entire East and Gulf coasts, considerable and valuable coastal acreage will be permanently submerged by sea level rise. Many small towns and cities will be lost.

But for places where local conditions aggravate sea level rise and flooding, the situation will be considerably worse. For example, the sea level has already risen nearly 12.5 inches at Galveston, Texas, 10.5 inches at Norfolk, Virginia, and 9.5 inches at Atlantic City, New Jersey, while rising six inches at Boston and 2.5 inches at Los Angeles.[18] Variations in ocean currents, characteristics of local geology, subsidence or uplift of landmasses, and exposure to storms are the reasons for these wide differences. New Jersey and Louisiana drew the short straws for most of these factors.

Variations in sea level rise along the Mid-Atlantic coast are affected by the strength of the Gulf Stream that appears to be weakening because of rising temperatures. Sea levels on the inner (westward) side of the Gulf Stream are several feet lower than on the outer eastern side. But this elevation gradient weakens when the Gulf Stream slows down and allows water to slosh up against the East Coast. This process will likely intensify with increased warming, meaning that the U.S. East Coast can expect to experience higher rises in sea level than the global average.[19]

Land elevation may rise or fall independent of sea level because of earthquakes, volcanic activity, and changes in continental mass due to melting glaciers or sedimentation. Land is rising in places that were covered with glaciers during the last ice age. In Scandinavia and northeastern Canada, uplift of the land due to this "glacial rebound" was an astonishing twenty-five feet per century while the ice was still melting ten thousand years ago. But rates have slowed to less than half an inch per year, which is roughly equal to global sea level rise. History helps put this in context. The original Viking settlement of Stockholm was a seaport located in a bay on the Baltic Sea. Passage into the bay became difficult about eight hundred years ago. Today Stockholm is on Lake Mälaren, which is still connected to the Baltic Sea through a system of locks that lift ships so they can pass into the lake, now two feet above sea level.[20]

Unfortunately, most of the planet does not experience uplift and may even undergo subsidence due to natural processes or human meddling. For example, most of the East Coast of the United States is subsiding due to the melting of the ice sheet that covered most of Canada during the last Ice Age. Compression by this gigantic, mile-high pile of ice pushed the land underneath it downward and caused an uplift of the surrounding land. But now that the ice sheet is gone, the land that was beneath the ice is rebounding upward, causing the surrounding land to subside. The hinge point between this upward and downward movement is in northern New England, with the greatest subsidence centered throughout North Carolina and the Chesapeake Bay.[21]

Some places, especially deltas, are sinking much faster than the sea level is rising. Nearly half a billion people live on deltas, typically in major cities. Local circumstances vary, but there are two principal reasons for this, both caused by people. Deltas subside because the natural mineral sediments that built and renewed them over millennia are being trapped behind dams upstream, and because the loam enriched by centuries of decayed plant material is becoming more compact due to oxidation and the extraction of water, petroleum, and natural gas. Compaction also affects inland areas subjected to extensive fracking, as recorded by the enormous increase in earthquakes in Oklahoma. Misguided coastal engineering can also be a major factor, as in the case of Mister Go in New Orleans.[22]

Subsidence can be severe and affect large areas. The Ganges Delta spanning Bangladesh and India is the largest in the world. It covers over 39,000 square miles and is home to 140 million people. By comparison, the Mississippi Delta covers just 4,600 square miles, with two million inhabitants. However, land areas less than six feet above sea level on the two deltas are quite similar, ranging from about 2,400 to 2,700 square miles. Where deltas are subsiding as sea levels rise, the sea level is rising six to eight times faster than the global average.[23]

The Mississippi Delta is disappearing before our eyes. One quarter of the wetlands are gone, and sedimentation is well below that required to keep up with subsidence. At current rates, the entire Mississippi Delta region will be submerged by the end of the century. The southern Chesapeake Bay region is also subsiding faster than sea level rise. This threatens major towns and the U.S. Navy base at Norfolk, Virginia. Spring floods are already commonplace, and the naval complex will almost certainly have to be moved at enormous expense.[24]

The composition of the land itself is also important. The hard, rocky coasts of much of New England are strongly resistant to erosion. In contrast, the entire U.S. seaboard from New Jersey to northern Florida, as well as most of the Gulf Coast, are sandy coastal plains easily eroded by storms.[25]

So far, we've discussed average sea level without regard to weather. But everybody knows that weather makes all the difference. We speak from personal experience. Steve flew into the eye of Hurricane Gustav on the Gulf Coast, and Jeremy has witnessed his fair share of extreme storms. As a child in 1948, he lived through two hurricanes in the old Hotel Floridian on Miami Beach, and witnessed many lesser gales in South Miami. But these were just the warmup for Hurricane Allen in August 1980, the fifth most intensive hurricane in recorded history with sustained winds of 186 miles per hour. Jeremy and his wife, Nancy Knowlton, were doing research at the Discovery Bay Marine Laboratory on the north coast of Jamaica when the storm hit. They rode out the hurricane with thirty colleagues in their rented house high in the hills, hoping the roof would hold. The waves were taller than trees. New islands were constructed overnight from the debris of reefs that had been utterly destroyed. Afterward they studied the storm's effect on the coral reefs.[26]

Hurricanes and typhoons are tropical cyclones. Low atmospheric pressure at the storm's center (the "eye of the storm") produces strong winds, rising seas, and flooding rains. Cyclones are heat engines that form over the oceans where sea surface temperatures reach at least 80 degrees Fahrenheit. Most dissipate at sea, but about one-fifth of them blow onto land. The strongest cyclones occur in the North Atlantic, where the wind speeds, as Jeremy witnessed, can exceed 180 miles per hour. The storms feed on the heat released when warm, moist air, fueled by evaporated seawater, rises from the sea surface before falling back as heavy rain.[27]

Hurricanes Katrina in 2005 and Sandy in 2012 were the first and fifth costliest natural disasters in modern U.S. history, respectively.[28] Public and scientific scrutiny focused on two major questions. First, are hurricanes becoming more frequent or more powerful because of global warming? Second, how does sea level rise contribute to the destructive power of hurricanes, regardless of the other effects of climate change?

Because cyclones are fueled by heat, one would expect global warming to generate more frequent and stronger storms. But because few long-term records exist and cyclone occurrence varies greatly from year to year, it's difficult to rigorously demonstrate cause and effect. Kerry Emanuel, a meteorologist at MIT, studies the link between global warming and cyclone intensity. His Power Dissipation Index (PDI) defines the destructive potential of a storm in terms of maximum sustained wind speed, which is easy to measure. PDI is simply the cube of the maximum sustained wind speed, and it relates to potential economic damage. We can calculate this on our phones. For example, Hurricane Allen's PDI was 6,434,856. Now, consider a hurricane with maximum sustained wind speed of 125 miles per hour. If its wind speed increases by 5 percent to 131 miles per hour, the PDI will increase by 16 percent. But if the maximum wind speed increases by 10 percent to 138 miles per hour, the PDI will increase by 33 percent, and for a 15 percent increase in wind speed, the PDI will increase by a whopping 52 percent. And so on. Thus, small increases in storm intensity can result in catastrophic increases in damages and loss of life.[29]

Emanuel demonstrated a significant correlation between his index and sea surface temperature. An increase of nearly one degree Fahrenheit since the 1930s, he explained, is fueling stronger storms. He has argued that this rapid rise since the industrialized twentieth century is strong evidence for a human cause, but critics said that it's too soon to rule out natural variability. Nevertheless, although the available time-series data are inconclusive, high-resolution dynamic climate models do predict a shift toward stronger storms due to global warming. For major storms in the future, these models project an increase in storm intensity from 2 to 11 percent, as well as a 20 percent increase in rainfall within sixty miles of the storm center.[30] Most recently, the extraordinarily rapid intensification of Hurricane Harvey leaves little doubt about the role of increasing temperature in fostering stronger storms. Harvey was a mere tropical depression when it encountered a patch of exceptionally warm water in the upper Gulf. Fueled by this jolt of heat, the storm intensified from a tropical depression to a category 4 hurricane in just fifty-two hours, faster than anything before.[31]

In addition to direct wind damage, wind-driven storm surges can exacerbate flooding and economic loss. The greatest threats are to broad stretches of shallow water extending inland from the coast—including bays, marshlands, and deltas. Coastal geometry is fundamentally important. The New York and New Jersey metropolitan area is particularly vulnerable to hurricane storm surge because it lies between the outlets of two enormous funnels: the New York and New Jersey Bight, and Long Island Sound (see maps in the Epilogue). Onshore winds from hurricanes like Superstorm Sandy push enormous amounts of water into the funnels that, as the funnels narrow, have nowhere to go but up. Long Island Sound, for example, is roughly fifteen miles wide near Rhode Island at its eastern mouth, but it narrows to the west, so that it is about two miles wide at La Guardia Airport, half a mile at Throg's Neck, about six hundred feet at the RFK (Triborough) Bridge, and one- to two-thirds of a mile along the East River between Manhattan and Brooklyn. The city is even more vulnerable to the south. Water pushed into the Bight is forced into three narrow channels at the entrance of Jamaica Bay, the Upper Bay of New York Harbor where it runs into the Hudson and East rivers, and the narrow channel called the Arthur Kill that separates Staten Island in New York from coastal New Jersey and then empties to the north into landlocked Newark Bay. The Great South Bay along the southwest coast of Long Island also has a funnel-like morphology. All in all, it's a landscape fine-tuned for disaster that was made even worse during Hurricane Sandy because the storm struck at high tide. Peak storm surges during Sandy reached ten feet with the worst flooding in the Great South Bay, Jamaica Bay, Arthur Kill, the East River, and westernmost Long Island Sound.[32]

The severe flooding damage to New York had actually been predicted three times before the storm. Two exhaustive studies published five and two years before Sandy concluded that, by 2100, slightly stronger hurricanes, combined with a sea level rise of about three feet, should mean that storm surges which formerly occurred only once in a century would occur much more frequently, every three to twenty years. A third comprehensive report on adaptation to climate change in New York State outlined in depth the dire risks that the city would face during a major hurricane—the same year that Hurricane Irene nearly drowned the subways and tunnels.[33]

Klaus Jacob, lead author of two of these warnings, tried to get New York officials to understand how vulnerable the city was to another strong hurricane like Irene that could trash the transportation and communications systems in a day. Transportation and communications are the lifeblood of the city's dominance in American finance, but nobody paid close enough attention. Then, with Sandy knocking on the door, Jacob was finally noticed; the *New York Times* called him the Cassandra of New York City subway flooding. After the storm, too, *Time* magazine singled him out as one of the "People Who Mattered in 2012."[34]

Storm surge had caused extensive damage in New York City before, when the sea level was lower and fewer buildings had gone up on landfill. The situation is more precarious now. Co-op City in the Bronx, for instance, is the largest cooperative development in the world, with 55,000 people packed into thirty-five high rises and other units. Like Venice, it is situated on a drained swamp. Tower foundations—fifty thousand pilings—extend down through the fill to bedrock. But the land is slowly sinking. Cracks are appearing in surrounding structures as the water seeks to reclaim its domain.[35]

Kerry Emanuel's PDI predicted that financial damages due to extreme storms should increase as the cube of the sustained wind speed. But actual damages are vastly worse because hurricane damage reflects not only the wind speed but also the quality of the construction where people choose to live. William Nordhaus is a distinguished Yale economist, member of the National Academy of Sciences, and an expert on the economics of global warming and climate change. He is particularly interested in how global warming will shape the fate of coastal cities. Two years before Hurricane Sandy, Nordhaus examined the storm characteristics and economic damages caused by 233 hurricanes since 1933, as well as thirty earlier storms. He found that storm damages increased as the ninth power of maximum wind speed, not the third power as proposed by Emanuel. To put this in more concrete terms, imagine that the damage to a city by a 150 mile per hour storm is $100 million. By the ninth power rule, damages due to a 160 mile an hour storm would increase to $167 million and for a 170 mile an hour storm damages would triple to $309 million. Nordhaus attributed this alarming difference from Emanuel's results to the compounding effects of storm size, storm surge, and rainfall in addition to wind speed, and the vulnerability of buildings and other infrastructure to catastrophic damage and failure. To illustrate his point, he made an analogy with the collapse of the World Trade Center on 9/11. All buildings and built infrastructure are constructed within predetermined tolerances. When the heat of the fires exceeded the buildings' structural limits, the Twin Towers collapsed.[36]

Nordhaus considered the future economic consequences of extreme events such as Hurricane Sandy in the context of stronger storms and sea level rise—existential events outside our range of previous experience. Under these circumstances, the normal economics of cost-benefit analysis and risk assessment might break down. Meeting with Jeremy over lunch, Nordhaus suggested that the best analogy might be total destruction in warfare. In Dresden, Hiroshima, and Nagasaki at the end of World War II, everything was lost and survivors had to rebuild from scratch. But the big difference in relation to sea level rise is that the land was still there. In southern Florida and coastal Louisiana, such an extreme storm would cause the land to disappear underwater, submerging the homes and hopes of more than 10 million Americans-turned-climate-refugees.[37]

Nordhaus described an exercise he gives his students at Yale.[38] The campus is in New Haven, Connecticut, roughly twenty-five feet or more above nearby Long Island Sound. The exercise asks students to imagine they are the president of the university, and presents them with three scenarios. The first is a tsunami with just thirty minutes of warning. The only response is to sound the alarm and run to high ground. Maybe grab your laptop but nothing else.

The second is a little far-fetched, but makes its point. Imagine that a small asteroid will hit the earth in ten years, probably in the ocean and almost certainly generating tsunamis fifty feet high. There will be plenty of time to move the art, the library, scientific instruments, and the other valuable stuff out of harm's way, then come back after the flood to see what's left, and whether it's worth repairing or it's time to move. This scenario might actually happen if we discover that the Antarctic or Greenland ice sheets have begun to break apart.

The third example is most relevant to gradual but accelerating climate change. It is highly likely that the ocean will be lapping against the campus in two hundred to three hundred years. The university, the city of New Haven, and all the cities and people on the Long Island shore will have to move. So the question becomes where and when? Nearby but well above maximum sea level rise? Or someplace entirely different with new opportunities yet to be determined?

If the migration toward the coasts continues at its present pace, half the population of the United States may have to make this kind of decision in the next century. Too many places like New York City, Miami, and New Orleans are a lot more vulnerable than the Yale campus. The question for all of us is whether we will get our act together and move well beforehand, or wait until we have to grab our laptops and run.

The U.S. Navy cares a lot about climate change for a great number of reasons, not least because many of its bases are gravely threatened by rising sea levels and it is the service most frequently mobilized to deal with coastal emergencies around the world. In January 2015, Jeremy gave a lecture at the U.S. Naval War College in Newport, Rhode Island.[1] It was about mounting security and humanitarian issues confronting the United States worldwide due to sea level rise, climate change, and environmental degradation. The audience consisted of War College students, faculty, and senior officers—including several admirals. Experts in national security, terrorism, strategic studies, and emergency planning heard his presentation with great interest. Except for a single climate change denier, participants in the lively discussion that followed accepted Jeremy's sobering conclusions as given and probed their implications for the future. It was his third such visit since 2009.

Fortunately the rest of the country is also catching on, and none too soon. High tides are getting higher, and flood tides are getting higher still. Hurricanes Katrina and Sandy were terrible, and Harvey, Irma, and Maria were all the more terrifying for striking so strongly in such rapid succession. In Florida and North Carolina, where Republican state legislatures banned any official mention of sea level rise in government venues, coastal residents worry about steep increases in federally subsidized flood insurance—if they are lucky enough to get it at all. And while trophy homes on Miami Beach continue to sell to people who can afford them, home values in much of south Florida are barely holding ground.[2] With risks and insurance rates going through the roof, everything points to getting out while the getting is good. But leaving homes, jobs, schools, and friends can be so overwhelmingly difficult that people instead soldier on and hope for the best.

People need help and responsible leadership if they are going to come to grips with the inevitable and react intelligently. Adaptation to the reality of a rising sea level requires three big steps. First is understanding and assimilating the scientific realities, which is more challenging than it should be because the public is increasingly uninformed and distrustful of basic science. Second is incorporating those broad realities into knowledge about particular situations. And third is acting on that knowledge to increase the safety and wellbeing of everyone affected. These three factors vary markedly from place to place because of differences in location and culture. For example, the sea level is rising much more quickly off the coast of New Orleans than near Miami or New York City because of the rapid subsidence of Louisiana's coastal lands. Cultural and political competence to deal with the situation is critical, and depends on education and access to accurate information. Meaningful action will in turn depend on a community's resources and quality of governance.

Let's make adaptation less abstract by considering the situation confronting three U.S. cities threatened by sea level rise: New York City, Miami, and New Orleans. In 2011, engineers and economists in association with the Organization for Economic Cooperation and Development (OECD) ranked these cities as the third, first, and twelfth most vulnerable worldwide to just twenty inches of sea level rise by 2070—an amount well within the last IPCC expectation but well below more recent estimates that take into account the increasing instability of the Greenland and Antarctic ice sheets. These rankings will change as coastal cities in Asia experience explosive economic and population growth, but as they are, they bring the risks closer to home.[3]

Only 7 percent of New York City would be flooded by five feet of sea level rise (with 20 percent of Jersey City inundated and Newark, New Jersey, just 2 percent underwater). But the entire New York metropolitan area is highly vulnerable to storm surge. Financial assets at risk today are about $320 billion, a sum that is projected to rise to about $2.2 trillion in present-day value by 2070, when a whopping 2.9 million people will be exposed to floods in low-lying areas.[4]

That said, New York City has a lot going for it. There is a growing understanding of the danger among ordinary citizens and private institutions and in state and city government. Rational discussions have begun. Museums are educating the public by mounting exhibits about restoring marshlands for greater coastal protection, and about retreating from vulnerable areas like the Rockaways and Battery Park. Well-constructed storm-surge barriers at the entrance to New York Harbor and the northern entrance to the East River, which are being discussed, could buy a century of protection. Other safeguards against flooding from the Hudson River due to torrential hurricane rains include placing seawalls at the most vulnerable points along the shore, capping and raising the entrances to the subways, and putting electrical and computer centers several floors up. The cost for the whole Cadillac package is a cool $35 billion, but that's only a little more than half of the $65 billion losses due to Hurricane Sandy and one-tenth of the economic assets that are exposed today.[5]

So far, these plans are mostly talk as experts and government officials dither and balk at the expense in the face of other needs like education and housing—although it could all be financed by a combination of bond issues, occupancy taxes on tourists, and other financial structures. As things stand at the time of publication, the city is still as vulnerable as it was when Hurricane Sandy struck. Nevertheless, given its extraordinary wealth and importance to the U.S. economy, its above-average governance, and the viability of fortifying much of its shoreline, New York City has a whole lot of options before it must begin to seriously consider vacating from the coast.[6]

Greater Miami is an entirely different story. Its average elevation is barely over three feet, with $416 billion in exposed assets today, a figure that will rise to $3.5 trillion in 2070, when the city will have an exposed population of 4.8 million. Throw in the rest of low-lying southern Florida, including Tampa, St. Petersburg, and Fort Lauderdale, and the numbers soar to something like $10 trillion in exposed assets and ten million people threatened by 2070. Moreover, most people don't understand the risks. There hasn't been a truly strong storm surge accompanied by heavy rains since the Great Miami Hurricane of September 18, 1926—a storm so severe that it ended the real estate bubble of 1925 and presaged the Great Depression.[7]

The storm was enormous. Miami was ripped apart by 145-mile-per-hour winds, and hurricane-force winds extended over 190 miles. A storm surge ten to fifteen feet high swept across Miami Beach and Biscayne Bay, and into downtown Miami. Knee-high water filled Miami and Miami Beach, and the MacArthur Causeway connecting the two cities was submerged under five feet of seawater. The Red Cross reported 372 deaths out of a city of only 111,000. That may be less than a third of 1 percent of the population, but if we apply that same ratio to today's greater Miami area population, which is about six million, that would be the equivalent of more than twenty thousand dead. Damages were estimated at $105 million in 1926 dollars, or roughly $164 billion today—$51 billion more than Hurricane Katrina.[8]

In 2017, history nearly repeated itself until Hurricane Irma took a Hail Mary last-minute turn on September 10 to cross over the Florida Keys and southwestern Florida instead of crashing directly into Miami. Irma also providentially veered into the central backbone of the state as it moved northward, sparing Tampa and St. Petersburg from the anticipated ten- to fifteen-foot storm surge that would have leveled the entire region. But despite being spared the worst, the damages were enormous. Miami and Miami Beach were severely flooded even though the eye of the storm was more than a hundred miles to the west, more than ten thousand people were left homeless in the Florida Keys, Gainesville and Jacksonville were severely flooded, six million people were without power for up to several days, and at least forty-two people died. Early damage estimates, which are bound to skyrocket as final numbers come in, are about $65 billion. And that was for a so-called miss. This is just an inkling of the horrific consequences of a direct hit to a major metropolitan area.

In response to explosive economic and population growth and in abysmal ignorance or flat denial of its history, housing in Miami has come to include dangerously exposed coastal areas like Key Biscayne and artificial islands built on landfill. No sane person would have built houses there in the 1920s. To make matters even worse, southern Florida's bedrock is highly porous limestone that takes up water like a sponge. The water dissolves the rock, creating large underground caverns that collapse as the sea level rises. The resulting massive sinkholes have gobbled up entire houses, sometimes with people inside. The ground is literally disappearing under people's feet.[9]

Yet despite these facts, the state of Florida and local governments have continued to collaborate with developers and the tourism industry to aggressively promote new construction on more and more landfill. Orrin Pilkey—a leading American expert on coastal exposure to sea level rise—calls Florida an "outlaw state. Florida has been particularly irresponsible and it's going to pay the price very soon."[10] Tragically, the ability of ordinary people to escape financial ruin seems to diminish every day.

Floridians' safety valve had always been federally sponsored flood insurance. For the mostly, but not always, well-off people who could take advantage of it, federal insurance underwrote houses on the Florida coast. Homeowners didn't worry too much about hurricanes and sea level rise because if disaster struck, the federal government would rebuild their homes in the same place at taxpayer expense. An astounding 35 percent of all houses insured through the National Flood Insurance Program are in Florida. But federal flood insurance must still pay off the $23 billion debt it accumulated in claims from Hurricanes Katrina and Sandy. Consequently, new insurance rates are skyrocketing. The mandatory insurance requirement for anyone with a federally backed mortgage—for example, from Fannie Mae—means those homeowners will face annual insurance bills of $25,000 or more. While the higher price more accurately reflects insurance risk, homeowners don't like it. They may no longer be able to afford their homes, they may no longer be able to sell their homes, and they blame the federal government for what in reality is everyone's responsibility.[11]

Private insurance companies have also played a major role in Florida. But eleven of thirty companies selling premiums in Florida went bankrupt after record-breaking losses caused by Hurricane Andrew in 1992. The remaining companies restricted future policies and raised their rates, causing one million Floridians to lose their insurance. At that point the state set up state-backed insurance programs that morphed in 2002 into the Citizens Property Insurance Corporation, which immediately became the biggest insurer in Florida. After even more record-breaking losses in 2004 and 2005, the private insurance industry largely withdrew from Florida because it was too risky. The state then became the major insurer, offering home insurance policies that it could not afford to pay off on in the event of a major hurricane. The subsequent administration has tried to reverse this untenable situation by allowing private insurers to calculate risk plus a reasonable profit based on the best data available. This allowed Citizens Property Insurance to shed more than one million policies to the private market, but that process is slowing down. As of December 2016, state-owned Citizens still held nearly half a million policies and that number is expected to increase next year.[12] Claims will obviously skyrocket in the aftermath of Irma, and it will be interesting to see whether the state-backed company will be able to pay them off.

Steve interviewed Robert Hartwig, president and chief economist of the Insurance Information Institute, to better understand the role of the private sector in Florida.[13] Hartwig believes that federal flood "insurance subsidies create perverse incentives for people to assume irrational risks. And this creates increased exposure for the U.S. taxpayer." He also believes that the market should set insurance rates based on objective assessments of variations in risk from place to place. This inevitably means that flood insurance in Florida and Louisiana should be sky high because of the "record-setting number of claims and cost of claims in the state."

Steve asked Hartwig how the insurance industry is dealing with climate change, specifically with the rising sea level, higher storm surges, and the likelihood of stronger and more damaging storms. He answered that insurers are keeping a very close eye on the volatility and the science of climate change and accurately assessing risks. "Insurers have been dealing with variability and volatility for centuries. If they weren't good at it, they would have all been out of business long ago. So under any condition of risk, insurers should be able to provide products that businesses and homeowners need. It's only a matter of being allowed, being able, to charge a risk-appropriate premium that allows a reasonable rate of return given the capital that has to be allocated. Insurers are well cognizant of the risk. They've modeled the risk, they've priced the risk, it's available on the market, and so the only obstacle here is insuring that people purchase the coverage."

Hartwig told Steve that only a tiny fraction of the property in Miami is actually insured against a direct hit by a major hurricane on the order of the 1926 storm. Hartwig put its insurance exposure at about $100 billion, while the total "insurance in force" for Florida on the FEMA website is $430 billion. Most people don't buy flood insurance, Hartwig asserted, because they believe the federal government will bail them out. "And so they rationalize and say, 'Well, why would I bother to buy the coverage if the federal government is going to provide me post-disaster aid?' But it's an unfortunate calculation. Some people are still waiting for aid from Sandy. And it won't be sufficient to help them completely recover."

You can't sell a house that's underwater, either literally or figuratively. Most of the uninsured assets in south Florida are in real estate, property that will likely be worthless after another direct hit by another "Great Miami hurricane." The environmental future of the southern two-thirds of Florida is bleak. It can either experience sudden, catastrophic destruction by another super hurricane, or slow death by submerging infrastructure as the sea level rises, with saltwater contaminating the drinking water, tidal flooding becoming a daily experience, and the ground collapsing into sinkholes. Miami Beach already experiences routine flooding.[14]

The United States generates more and better climate data than anywhere else in the world, yet we are painfully slow to act on the basis of what we know. In contrast, Cuba has already conducted a rigorous, nationwide scientific assessment of the risks of powerful hurricanes and storm surges associated with rising sea levels. Using that scientific assessment, Cuban leaders constructed a risk map and zoning plan for their entire coastline, and enacted laws to empower the government to act on that information. They are protecting their citizens by relocating dwellings and entire towns inland to higher ground, to the extent the government can afford to. The Cuban government has also enacted a strongly proactive approach to disaster preparedness, with well-developed educational programs designed to prepare the populace for emergencies.[15] It is deeply ironic that a somewhat impoverished Cuba is facing the dangers of sea level rise realistically, and with significant resources, while Americans just a hundred miles to the north are still arguing about whether the dangers are real.

Proven engineering solutions make sense in places like New York City, which is built mostly on solid bedrock with adequate elevation, although these are expensive fixes. Klaus Jacob, the famous Cassandra of the New York subways, said in an interview that there are three fundamental ways to adapt to climate change and sea level rise. The first is to protect: keep the water out with dams and barriers like the one at the mouth of the Thames Estuary that protects London (a similar solution is being discussed for New York City). The second is to accommodate, by letting the water enter in strategic places in order to increase the security of protected land. This is the strategy the Dutch are pursuing so effectively in the Netherlands. The third option is strategic resettlement, which could be highly effective in New York City where there are locations as high as 220 feet above sea level nearby. But in Miami, even resettlement is hopeless. The highest elevation is eighteen feet above sea level, and the city "sits on sponge-like limestone, which is worse than Swiss cheese: the holes are all connected to each other. If you build a dam or levee on it, the water just flows underneath."[16]

We sadly agree. Massive inundation and saltwater intrusion are inevitable throughout greater Miami—either gradually over the next few decades or anytime after the next direct hit by a Category 4 or 5 hurricane with high storm surge, for which the city is poorly prepared at best. There is a quite good Miami and Dade County hurricane readiness guide with a simple-to-understand map of areas most vulnerable to storm surge. Incredibly, however, a search on the web failed to discover any detailed and widely, publicly announced emergency plans for evacuating millions of people in a single day. Nor did we find evidence of widespread signposting of evacuation routes like the "Turn around and don't drown" warning system for flash floods in the canyons west of Boulder, Colorado, or the tsunami evacuation signs all along the California coast. Nor the kinds of mandatory hurricane education programs that all Cuban children complete. But then again, if the U.S. government officially denies climate change and a rising sea level, why would they bother to remind its citizens that they live in imminent danger?[17]

The challenges for evacuating hundreds of thousands of residents and tourists off of Miami Beach and Key Biscayne across narrow causeways to the mainland are huge. The challenges of evacuating six million people from Greater Miami trying to drive north all at once on I-95 are even greater. The near miss of Hurricane Matthew was an eye-opener. Governor Rick Scott showed real leadership in issuing dire warnings and mobilizing the National Guard, for which he was criticized afterward when the state, miraculously, escaped severe damage. And he did an even better job of getting people out of harm's way before Irma. But despite his warnings and the impressively rapid setting up of shelters for people who could not evacuate, Florida was not sufficiently ready and a direct hit would have been a monumental national disaster. Greater Miami has ten times more people than in New Orleans when Katrina hit, and we recoiled from the scenes of chaos and death that happened there. Another Great Miami Hurricane would create six million climate refugees and cause the financial collapse of Florida. It would be our Fukushima, for which we are just as badly unprepared.[18]

Meanwhile, as Governor Scott continues to question the reality of sea level rise due to climate change, the more realistic leaders of

Miami Beach are doing everything in their power to stave off the inevitable for as long as they can.[19]

Moving westward, the average elevation of New Orleans is about two feet below sea level with the highest elevations—along the levees that hold back the Mississippi River, which flows well above most of the city's streets—measuring fifteen to twenty feet above sea level. Large parts of the city are subsiding at the extraordinary rate of one to two inches a year. And not just in New Orleans: the entire Mississippi Delta region is disappearing underwater at the rate of a football field area of land every hour, the equivalent of sixteen square miles a year. That works out to about 1 percent of the entire Delta every year. This land is home to half of the oil refineries of the United States, 90 percent of its offshore oil production, and 30 percent of the U.S. oil and gas supply. The oil infrastructure in Texas around Galveston and Houston produces another 25 percent of the U.S. oil and gas supply and is just as vulnerable—as was tragically demonstrated by Hurricane Harvey barely two weeks before Hurricane Irma. All of this is poised for an economic and environmental disaster from the next direct hit of a hurricane on the scale of Katrina or Matthew.[20]

New Orleans today has 344,000 inhabitants, but in 2012 the combined metropolitan statistical area was home to more than 1.2 million people. Everyone in the region is critically threatened by the rise in sea level and the next great storm. Exposed assets add up to about $234 billion today, a figure that is projected to rise to about $1 trillion by 2070. These estimates do not include the vast petrochemical plants that line the 84-mile stretch of the river between New Orleans and Baton Rouge dubbed "Cancer Alley" for its pervasive pollution.[21]

Unlike Miami, most New Orleans locals understand their situation. The state of Louisiana, too, has finally admitted its extreme vulnerability; the governor has declared his intention to declare coastal erosion a "state emergency" and the release of the newly ratified Coastal Master Plan. Roughly half of New Orleans residents never returned after Hurricane Katrina, but as the population rebuilds with outsiders, the newcomers may not be so wise. Filling all the channels dug through the marshes for oil and gas pipelines might slow down the inevitable. Army Corps of Engineers levee projects might also help, at the cost of many billions of dollars. But with dubious success and ample evidence that the Corps made things worse, few locals have much faith in it. Indeed, the latest version of the Louisiana Coastal Master Plan paints a very grim picture of the chances for saving much if anything of the coast.[22]

One possible outcome for New Orleans would be to gradually allow the Mississippi River to run its natural course down the Atchafalaya River by bypassing the Old River Control Structure. The original dam there almost broke in 1937 and was entirely rebuilt upstream in 1967 as the enormously fortified structure it is today. Another failure would be a calamity, wiping out a vast area to the west and drowning forever the towns and people who live there. It would also turn the ports of New Orleans and Baton Rouge, as well as Cancer Alley, into a salty backwater, wiping out in an instant the entire oil and gas infrastructure of Louisiana. A breach in the dam might also provide some longer-term protection to New Orleans, although the Army Corps is determined to do everything possible to prevent such an event.[23]

According to Rob Dunbar they're bound to fail.[24] "The whole river's going to go where it wants to go; man's efforts are puny compared to it. Imagine flooding like we've had over the last two weeks, but on steroids over the last two months, that's when the Mississippi overflows its banks everywhere and, of course, the ability of floodplain areas to absorb it will be overcome and it will go downstream. Projects like Old Man River were built ages ago to keep the Mississippi out of the Atchafalaya, and it's been wanting to flow into the Atchafalaya Basin for a long time . . . It's crazy to think that this ancient manmade structure can't be removed by erosion or . . . like what happened in our reservoir in California, you look at those pictures [the Oroville Dam spillway failure that caused the evacuation of 190,000 people during 2017 flood rains]. That's a perfect paradigm for the Mississippi."

Dunbar continues: "They rebuilt some of the pumps and they raised the levees in a few key places, but when they do that they prepare for whatever they consider to be a ten- or a fifty-year [event . . . In] this changing world you've just got to toss these things out. There was never a good reason for defining a hundred-year flood event and building structures that could prevent this. Even when that concept was developed, it wasn't a fair representation of the risk associated with nature's extreme events. It was a shortcut that people could wrap their heads around, but [such approaches are] really off the table now as we're moving into this changing system."

The obvious but socially traumatic solution is what's barely mentioned in the Coastal Master Plan: Klaus Jacob's strategic resettlement, which in the case of Metropolitan New Orleans would mean moving entire, hopefully socially intact, communities far north to safety. It bears repeating that New Orleans and the Delta could be maintained as a transfer station for shipping and industry, manned by workers who could be helicoptered to safety on short notice. There are no serious plans for such relocation, however, and nobody wants to move—even if it will all happen anyway whenever another Hurricane Katrina hits New Orleans, which could be very soon. The costs of rebuilding yet again would be untenable. Moreover, the myth that all is okay is placing citizens in great peril.

The dilemma of sea level rise is also afflicting, on a smaller but equally meaningful scale, communities all along the East and Gulf coasts of the United States: Portland, Maine; Cambridge, Boston, and Cape Cod, Massachusetts; all of coastal New Jersey; downtown Baltimore and Washington, D.C.; the Delmarva Peninsula; all along the coasts of Virginia, the Carolinas, and Georgia, including Norfolk, Charleston, and Savannah: most of the Florida Panhandle; coastal Alabama and Mississippi; and Galveston and Houston. Clearly these cities and their citizens cannot face these challenges alone.

It's obvious from the soil loss, nitrogen poisoning, and super-weeds in Iowa and the Corn Belt that grain farming is in trouble. It's fixable trouble and we already have the know-how to do the fixing. But old ways die hard, especially for those heavily invested in the status quo.

Climate change makes all these problems worse and harder to solve, and it makes farming riskier. Agriculture accounts for a staggering 80 percent of all water consumption in the United States, more than 90 percent in the seventeen Western states. Therefore anything that affects water availability is critical to farming. Public attention has been focused on the massive drought that has desiccated California and the Southwest for several years. But in 2016, Alabama, Georgia, Pennsylvania, and other Eastern and Southern states suffered severe droughts, crop loss, and forest fires. New England corn died in the fields. And the rain that relieved these droughts often came in downpours that caused additional destruction.[1]

California had been dealing with drought for years until the deluge and severe flooding in the winter of 2016–2017. The most populous state in 2016 with 39 million people, California has eleven million more people than second-ranked Texas. But both states have had huge water problems for a long time. Population growth in California has slowed dramatically, from about 3 percent per year from 1900 to 2004 to just 0.7 percent today—the eleventh year in a row with less than 1 percent growth. Nevertheless, the drought has escalated the longstanding conflict over water between agriculture and cities, the two biggest contenders for this resource. Whenever that happens, cities tend to win. Remember the movie *Chinatown*, acclaimed as one of the best movies of all time? It was loosely based on 1930s Los Angeles. When power politics made the Santa Clara River flow uphill to steal water for the thirsty city, many Owens Valley farmers were left high and dry.[2]

Yet people are still coming to the Southwest in droves. Texas is growing at 1.6 percent a year, the tenth highest rate in the nation, and the flow shows no signs of slowing. Three of the top ten U.S. cities with over a million people are in Texas, a state buoyed by its booming—and sometimes busting—energy economy. Houston is the largest with a population of 2.2 million and the fourth greatest number of residents nationwide. San Antonio is seventh and Dallas is ninth. But the growth of metropolitan areas is even greater. The metropolitan statistical area of Dallas/Fort Worth/Arlington is ranked fourth in the United States with 7.3 million residents and Houston/ The Woodlands/Sugar Land is fifth, with 6.7 million. Houston, San Antonio, and Dallas are growing at rates between 1.6 to 1.8 percent. Austin is the eleventh-ranked U.S. city for population at just under a million residents and it has an annual growth rate of 2.9 percent, the highest for any city in the United States.[3]

All four of these largest Texas cities have serious water problems. Fort Worth is ranked nationally as the sixth most likely city to run out of water, San Antonio is ranked fourth, and Houston is second, topped only by Los Angeles in a competition that no one wants to win. People rightly obsess about the California drought, but Texas is drier.[4]

California also has three of the largest cities in the United States. Los Angeles with 3.9 million is the second most populous in the nation, San Diego is eighth, and San Jose comes in at tenth. Los Angeles is growing at only 0.8 percent per year, but San Diego and San Jose are increasing faster, at 1.6 and 1.2 percent, respectively. California's diversified economy is booming, and that prosperity is likely to continue. The tech, biotech, defense, and medical advances that come from the state's businesses, industries, and research centers, as well as the intellectual, social, and cultural capital spun by its entertainment, media, and social networking establishments, make California the sixth largest economy in the world, surpassing France in 2016.[5]

California also produces a huge amount of food and continues to lead the nation with $47 billion in sales in 2015, down from $57 billion in 2014 because of the drought. Sales include roughly half of the fresh vegetables and two-thirds of the fruits and nuts that Americans consume, plus exports. In the state's overall $2 trillion economy, however, food has always ranked near the bottom of the list, way below services, finance, real estate, manufacturing, technologies, communications, and several other sectors. Google and Apple combined approached $1 trillion in net worth in 2015, with annual revenues of $75 and $234 billion, respectively.[6]

California gets its water from a variety of sources that have been increasingly networked and centralized due to the drought. Most of its water comes from rain and snow from the northern and eastern parts of the state, but southern California, from Los Angeles to San Diego, receives most of its water from the Colorado River. All the sources are at or near capacity. Increasingly restrictive conservation measures are here to stay. But for all its water problems, California is better off than Arizona and Nevada, which also rely on the Colorado River system. The Colorado River provides water for seven states through an agreement signed in 1922 that awards the lion's share to California. But Colorado River water never seems to be enough. In 2016 the river's two major reservoirs, Lake Mead and Lake Powell, were down 64 percent and 54 percent, respectively, with barely a blip of relief from the extremely strong El Niño winding down at that time. With an ocean nearby, southern California is flirting with desalinating seawater on a massive scale, as the Israelis have been doing for half a century at a smaller level.[7]

In contrast, Arizona gets about 40 percent of its water from the same depleted Colorado system, and 40 percent from groundwater reservoirs that have been overtaxed for decades, particularly during very dry years. A remarkable 70 percent of that water is used to grow food in a desert, 8 percent is used for industry, and 22 percent goes to municipalities like Phoenix and Tucson. To make matters worse, a new study by NASA and the University of California at Irvine revealed that since 2004 the groundwater has been rapidly depleting:

three-quarters of the amount of freshwater that accumulates in the entire Colorado River watershed every year has been lost underground. This means annual groundwater loss far exceeds the annual depletion of Lakes Powell and Mead.[8]

So where will the water come from a decade from now? Groundwater resources are way down, and reserves may be good for only a few more years with current use. After the groundwater is depleted, agriculture will suffer first because the cities have priority. Cutting agricultural use can buy Arizona a lot of time. But if Lake Mead continues to dry up, even the cities will soon suffer—cities that, from the perspective of environmental sustainability, shouldn't be there in the first place.

Setting aside the obvious threats to several million inhabitants, Southwestern agriculture won't have a chance against the ongoing megadrought, which is predicted with some certainty to increase in severity if greenhouse gasses continue to rise.[9] The United States will face an enormous agricultural adjustment as Southwestern farmers figure out what can be grown where and how, and as markets decide whether that produce can be grown competitively. Limited primarily by water, California has always been the best place in the United States to grow fruits and vegetables on a massive scale. But what will survive in the great climate change lottery is up for grabs.

At least three challenges loom for American agriculture. The first is the impact of climate change on rainfall, temperatures, and the frequency of extreme weather events that will affect the continent in different ways in different places. The Southwestern drought and increasingly frequent extreme weather events are the most obvious impact of global warming, and they have grave implications for American agriculture as well as public safety. Extreme heat will likely become more severe, and records set for the hottest month and hottest day of the year will become more common. Although less probable, record dry years and consecutive five-day wet periods may also happen more frequently. Add to these the increased occurrence of severe hurricanes, outbreaks of tornadoes, and severe thunderstorms like the event that hit John Weber's farm, and you have a recipe for agricultural disaster.[10] Climate models and data over the past two decades provide strong hints about how all these issues may settle out, but there is still great uncertainty. Lives and fortunes will depend upon the outcome.

The second challenge is how the country will compensate for the likely reduction in fruits, vegetables, and nuts from California as water supplies destabilize and rationing becomes permanent. To some extent, new regions of the country may make up the difference. But California is the key. All bets are off until its water problems are resolved.

The third challenge is how quickly the Corn Belt can shake off environmentally destructive practices and produce grains, livestock, and bioenergy more sustainably.

Let's look more closely at the Southwestern drought. Make no mistake, the drought is real, the 2017 California flood rains notwithstanding. Its severity is unprecedented, and it increasingly appears to be the new normal. How do we know this? One resource guiding policymakers is a 2015 paper published in *Science Advances*. Written by Benjamin Cook and other scientists from NASA Goddard, Columbia University, and Cornell University, the title itself captures the problem: "Unprecedented 21st Century Drought Risk in the American Southwest and Central Plains."[11]

Extreme drought has played a major role in the collapse of civilizations. Ongoing warfare and the emigration crisis in rural areas of the Middle East and North Africa are largely due to enduring drought in the eastern Mediterranean that has led to social and political chaos for both the refugees and the societies that receive, or don't receive, them. At Mesa Verde and Chaco Canyon in the Four Corners country of Colorado, Utah, Arizona, and New Mexico, another severe, long-lasting drought about a thousand years ago caused these complex civilizations to disintegrate entirely. Similar events are widely believed to have caused the collapse of Mayan civilizations. The ensuing wars, starvation, and cannibalism were horrific.[12]

Apocalypse aside, Benjamin Cook's paper is deeply disturbing. For the Southwest, sixteen of seventeen climate models predict the soil one foot below ground to become significantly drier than it was when Mesa Verde and Chaco Canyon collapsed. Almost as many models offer a similarly dire prognosis for the Central Plains. Risk of drought in the models rose to more than 85 percent for the second half of the century—twice the risk as that which held sway from 1950 to 2000 (a period for which the models agree exceptionally well with actual historical conditions). Ominously, the risk of extreme droughts lasting for many decades rose to about 80 percent after 2050 compared to less than 12 percent in both regions before 2000. The authors conclude: "Our results point to a remarkably drier future that falls outside the contemporary experience of natural and human systems in western North America, conditions that may present a substantial challenge to adaptation."[13]

The rainfall picture changes as we move eastward, but is equally alarming. Rain and snow throughout the Midwestern and Eastern portions of the country are predicted to increasingly concentrate in extreme weather events as powerful as the top 1 percent of storms today. In fact, occasional extreme rainfall events during long periods of severe drought have already increased by 37 percent in the Heartland, 27 percent in the Southeast, and a remarkable 71 percent in the Northeastern United States. Snowmageddon in 2010, Superstorm Sandy in 2012, and the Snowzilla storms in 2016 caused massive coastal flooding in places along low-lying coasts from the Carolinas to Maine. For Atlantic City, New Jersey, coastal flooding from Snowzilla was almost as bad as from Hurricane Sandy. And that was just a very bad snowstorm.[14]

Such huge storms can dump massive amounts of water on Midwestern fields in just one or two days. Deep gullies can form overnight, causing massive erosion that pulls away topsoil at the rate of about four tenths of an inch per year. Storms from April to June 2014, for instance, washed away 15 million tons of Iowa topsoil. Amounts of fertilizer in local aquifers spiked. Similar storms contributed to the massive cyanobacterial blooms in 2014 that poisoned Toledo's drinking water as well as other aquifers throughout the Corn Belt. The intervening droughts are also costly, with great year-to-year variations in corn and soybean yields and damage to soils. For example, only 10 percent of Iowa fields experienced severe drought conditions in 2011, but 100 percent experienced extreme drought in 2012.[15]

Higher temperatures are also a major problem for our agricultural future, and heat needs to be considered separately from drought. A heated planet creates rain as clouds rise off warming oceans and condense at higher altitudes. But rain is becoming increasingly more sporadic, and dry times between downpours are getting longer and hotter. This is bad for corn and soybeans in the Heartland, as Wolfram Schlenker and Michael Roberts document in the *Proceedings of the National Academy of Sciences*. They compared yields of corn, soy, and cotton with fine-scale weather data for individual counties, which included the maximum distribution of temperatures each day over the growing season at specific locations. Yields increased up to 84 degrees Fahrenheit for corn, 86 degrees for soybeans, and 90 degrees for cotton, but temperatures above those limits were harmful. Modeling the results across current growing regions, by 2100, yields are predicted to decrease by 30 to 46 percent under the slowest warming scenario, and by a startling 63 to 82 percent under the most rapid warming scenario. The effects of rising temperatures on wheat production are just as severe, with productivity predicted to fall slightly more than 6 percent for every two-degree (Fahrenheit) increase in temperature.[16]

So why not just move the cornfields farther north? That's already happening. In 2013, southern Alberta, Saskatchewan, and Manitoba corn production was up by 78 percent and farmland prices rose by 12 to 47 percent. Prices and production have mostly continued to rise. That's bad news for the Heartland. Wheat farming is also moving northward and global production is declining with rising temperatures. How far north wheat can be farmed may depend on the distribution of suitable soil, since most of northern Canada's soil was scraped off by Ice Age glaciers and carried southward to be dumped on rich prairie lands in southern Canada and the U.S. Midwest. Consequently, northern Canada is mostly rocks.[17]

It is true that increasing levels of carbon dioxide in the atmosphere may be really good for plants. That's because carbon dioxide and water are the raw materials for producing sugar by photosynthesis, which is essential for plant growth. Other things being equal, increasing carbon dioxide might therefore offset the harmful impacts of increased drought and extreme warmth. But once again the situation is more complicated. Different kinds of plants respond differently to carbon dioxide enrichment because of their particular biochemical pathways for photosynthesis. The most important groups are referred to as C3 and C4 plants. Soybeans are C3 plants, which should benefit mightily from increased carbon dioxide. But that advantage is offset by their lower tolerance to the warmer and drier climates across the Heartland due to climate change as well as increased competition from superweeds. Corn and other C4 plants derive little benefit from high carbon dioxide levels, and this benefit is more than offset by the effects of increasingly severe droughts.[18]

The Northeast also faces great challenges due to climate change, principally because of desiccating droughts interspersed with unpredictable episodes of flooding caused by torrential rains. The already longer and warmer growing seasons may help some crops, but they come with periods of extreme heat. Temperatures above a species' tolerance level cause harmful stress. Corn's yield decreases by 2 percent for every growing day above 86 degrees Fahrenheit. Tomato and milk production are also reduced when temperatures soar. Maple syrup production has already declined drastically. In the 1970s, 80 percent of maple syrup came from the United States and only 20 percent from Canada. Now these proportions have flipped, with only 20 percent of total production occurring in the United States. Other crops that require long, cold winters—like cranberries, wild blueberries, many apple varieties, and Concord grapes—will inevitably decline, especially in southern areas of the region.

Warmer winters also allow more insect pests and weeds to overwinter and proliferate in the spring. Plant diseases, parasites, and invasive species from the south pose great threats to agriculture, and to forest products. As they march northward, two introduced insect species, hemlock wooly adelgids and emerald ash borers, have the power to change northern forests by destroying these native American trees. Confronting this perfect storm of threats requires rethinking what to plant where, and developing temperature-tolerant varieties of the traditional crops we rely on.[19]

Finally and inextricably, the crisis in American agriculture is linked with human health. If big agriculture is troubling, human health is its most obvious manifestation. Obesity is a national crisis, and the United States is the most obese nation in the developed world. More than two-thirds of Americans are overweight and a staggering 35 percent are medically obese. More than 190 billion dollars are spent on obesity-related health issues each year and that figure is sure to grow as people who have been severely overweight all their lives grow older. The decline in American health has been linked to an increasingly unhealthy modern American diet saturated with corn and wheat starch, high-fructose corn syrup, and red meat, major products of industrial agriculture in the Heartland.[20]

Americans' lives are significantly shorter than those of people in other wealthy nations. Life expectancy at birth is now seventy-eight years, which may seem like a long time, until you realize that in 2013 the World Health Organization ranked the United States thirty-fourth in life expectancy out of 194 nations, tied with Costa Rica and Qatar. In contrast, overall life expectancy in top-rated Japan was eighty-four years, and Spain, Italy, Switzerland, and Australia tied for second at eighty-three years. Cheese-and-wine-guzzling France tied with Canada and nine other nations at ninth with an average life expectancy of eighty-two years. Moreover, despite extraordinary advances in medicine, U.S. life expectancy is barely increasing, whereas life expectancies in most other high-income countries have been steadily improving. As we gain weight, we lose ground.[21]

Many socioeconomic factors have been examined to explain the stagnation of U.S. life expectancy. But obesity and related disorders—including diabetes, heart disease, and many cancers—are clearly a function of unhealthy diets.

Rising temperatures, the Southwestern drought, and increases in extreme weather events throughout the Midwest to the Gulf and East coasts are here to stay regardless of the encouraging progress being made in reducing global emissions below earlier "business as usual" projections.[22] American agriculture can and will adjust, but doing so will involve massive shifts in our decisions about what we should grow, and where we should grow it. Harmful agricultural practices compound the problems of climate change, but much can be done, including abandoning GMO crops dependent on Roundup. Fortunately, there are excellent alternatives.

There isn't just one big problem with agriculture, and when it comes to solutions, one size doesn't fit all. Each region's problems are different, and they all demand regional solutions. Climate models give us an increasingly good idea of the trends from region to region, but we still can't predict important details. A common aphorism attributed to Mark Twain says it best: "Climate is what we expect, but weather is what we get." For example, hurricanes are getting stronger but we don't know when and how hard the next storm will strike New Orleans or New York. We know that the California drought is almost certainly here to stay, but we can't predict temporary respites like the amazingly wet January 2017. Too bad California didn't have the infrastructure in place to keep that glut of water from running into the sea.[1]

Hurricanes and droughts are expensive. From 1980 to 2016, thirty-five hurricanes cost the United States over $500 billion in storm damages. Perhaps more surprising to many, the twenty-four droughts during this same period cost nearly $250 billion in agricultural losses. This means that the average drought cost the United States three-quarters as much ($10 billion) as the average hurricane ($14 billion).[2] Farming practices and regulations will need to be nimble and adaptive to respond to evolving conditions. But unlike the headlines regarding sea level rise and stronger hurricanes, there's a lot of good news for agriculture. Real, practical, and achievable solutions can actually fix most of America's agricultural problems, despite the inevitable uncertainties.

For California and the Southwest, the overriding problem has always been drought, just as Marc Reisner described in *Cadillac Desert*. The solution in the past for big cities such as Los Angeles was to take water away from other places, victimizing people who were powerless to defend their own interests. But there's no more water left in other places. Solutions now require a trinity of pain, sacrifice, and money.[3]

Like California, the Heartland increasingly suffers from wildly variable weather: too little water during droughts, then destructive rainstorms that wash away fields and damage buildings and equipment. Weather that flip-flops from one extreme to another, from drought to flood and back again, makes all other problems worse. But the Midwest's gravest problem, the chemical arsenal that supports industrial agriculture, is the most straightforward to fix if farmers want to do it. They can abandon Roundup and its noxious cohort of additives, and adopt proven and profitable farming methods known collectively as ecological or regenerative agriculture, even if so far industrial agribusiness has systematically ignored and actively campaigned against these strategies.[4]

In the Northeast, farms were largely abandoned by the 1950s as California increasingly came to dominate produce markets, and as Midwestern dairy products undercut Eastern production of milk, butter, and cheese. Fields and pastures cultivated for more than a century reverted to scrub and forest. But as cities have grown and markets for local and organic food have expanded, fallow land has been returning to small farm production. Local fruits, vegetables, and dairy products may soon satisfy much if not all of the East Coast's needs. Yet frequent droughts and floods also afflict the East, and make small-scale agriculture precarious.[5]

In typical fashion, California leads the pack in confronting uncomfortable realities. Governor Jerry Brown has become a hard-core, well-informed pragmatist pushing ahead to conserve and centralize scarce water resources statewide, and to encourage more innovative sources of water production. He quickly integrated stringent water conservation measures across all domestic and agricultural sectors after the severity of the drought became apparent in 2015. Less than a year later, after a much-heralded El Niño failed to break the drought, he made all water conservation measures permanent and warned of even greater restrictions ahead. Under intense pressure, he back-pedaled almost immediately to allow municipalities to manage their nonagricultural usage. Then, faced with continuing bad news about the severity of the drought, Brown mandated stricter, long-term measures that appear to be here to stay.[6]

Cities like San Diego have enacted sophisticated environmental plans across the board, and are beginning to implement them. There has been serious talk about replacing water-guzzling lawns with native vegetation that can survive on occasional rains. Residents and businesses that persist in consuming too much water may be summoned to court or publicly shamed into doing the right thing. But whether the agricultural empire remains intact will require a lot more than rationing and homeowner conservation. A lot more water must come from somewhere.[7]

One option is to follow the example of Israel, given that tiny country's remarkable domestic and export agricultural production (think Haifa oranges grown in the middle of a desert), which has been developed from scratch over the past seventy years. Israel is a tiny country, one-twentieth the size of California with only a fifth as many people. Nevertheless, the process of adapting Israel's water-shortage solutions has already begun.[8] The largest desalination plant in the Western Hemisphere is now operating in Carlsbad, a high-tech city in northern San Diego County. The Carlsbad Desalination Project was planned to supply about one-third of all the water consumed in San Diego County, with the goal of providing a stable and reliable alternative to water from the Colorado River and Northern California. The plant so far provides 50 million gallons of water a day, or about 10 percent of San Diego County's total demand. But it cost a cool billion dollars and still faces environmental and regulatory concerns.[9]

Reverse osmosis—pushing saltwater through a porous membrane with holes so small the salt can't get through—is expensive. The Carlsbad plant is the first of seventeen such facilities projected for the entire state.[10] Trade-offs will have to be made. Agriculture? Homes?

Biotech? How much will pipelines cost that must stretch across mountains and valleys to reach thirsty users, especially big agriculture, which is situated closer to the desert than the coast? What about earthquakes?

It takes a lot of electricity to push saltwater through filtering membranes, although to put things into perspective, the eight kilowatt hours required to provide one day's freshwater for an average household is only two-thirds of the power used by that family's water heater or electric car, and only one-third of the amount employed for central air conditioning. Since electricity costs money, the cost of running the plants will be determined by the price of electricity, and by how it's produced and supplied. In 2016, oil and gas hit rock-bottom prices, but how long that will last no one knows. Two-thirds of the electricity produced in San Diego is fueled by natural gas, followed by wind and solar at 15 percent and 4 percent, respectively. But if oil prices rebound, will California redouble its efforts to invest in more wind, solar, and other renewables to meet its needs?[11]

Better technologies and infrastructure are essential to help solve California's water crisis. But California cannot build its way out of its water problems without a major push for improved water conservation. Essential steps include greater efficiency of water use, reuse and recycling of water, and improved capture of stormwater to recharge groundwater levels and for non-potable uses such as gardens and agriculture. Structural improvements will include more reservoirs to hold the increasingly sporadic runoff of snowmelt from the Sierras and the torrential El Niño rains that cause floods and mudslides. Most of the rain that caused the massive flooding in January 2017 fell on the coastal ranges rather than in the Sierras, so new water catchments and reservoirs will need to be constructed to keep it from running out to sea. Cement boondoggles like the Los Angeles River that rush precious rains down a sluice and out into the Pacific will have to be demolished, rebottomed with cisterns, and banked with plants to better absorb the water that California can no longer afford to waste. The network of open irrigation ditches running up and down the Central Valley loses a third of its water to evaporation. Reconfiguring irrigation as a drip system would also save huge amounts of water and increase the yields of some crops.[12]

Much of this is already happening. But it will cost tens if not hundreds of billions of dollars to construct new dams and reservoirs to compensate for the decreasing volume of the once mightier snow pack of the high Sierras due to climate change. Preliminary estimates just to repair the Oroville Dam spillway, for instance, are half a billion dollars.[13]

Even so, thirsty food crops like rice, almonds, and pistachios may have to be grown in wetter, but still sunny places, and replaced by more drought-tolerant crops. The biggest water hogs of all are the pasture and alfalfa raised to feed cattle and dairy cows. Ultimately, the markets will decide what can be grown efficiently and profitably by a retooled, water-conscious Californian agricultural system, in comparison with other regions. It's hard to imagine anywhere else competing with the extraordinary volume and quality of California tomato farming, which Mark Bittman described so enthusiastically in the *New York Times*, and tomatoes actually require only tiny amounts of water compared to most other crops. Their profitability in terms of water use is also exceptionally high, like it is for vineyard grapes and the much-maligned almonds and pistachios. But waterguzzling rice gives a paltry dollar return and clearly has to go somewhere wet.[14]

Rising temperatures are also going to require a massive retooling of premium wine production, which requires a delicate balance of adequate heat accumulation, low risk of severe frost damage, and, most importantly, an absence of extreme heat in the growing season. If temperatures rise as much as predicted, the areas suitable for producing the highest quality wines will be reduced by 50 percent.[15]

Moving eastward into the Heartland, climate change will cause a wide range of problems described earlier: accelerated soil loss, excessive use and runoff of pesticides and nitrate fertilizers, toxic algal blooms, groundwater and reservoir contamination, explosions of superweeds, downstream pollution and dead zones, the collapse of pollinators, a loss of biodiversity, and increasing risks to human health.

We know right now how to fix all of these problems sustainably, without losing production or profits. Taxpayers have funded the necessary research for over twenty years. Scientifically rigorous and repeatable large-scale agricultural experiments have yielded simple twofold solutions. The first is to increase the diversity of crops grown in rotation on the same field, which drastically reduces the need for fertilizers, pesticides, and water. The second is to use a mix of annual and perennial species planted in buffer strips to reduce surface runoff of nutrients and soil, as well as catchments to remove nutrients from tile drainage before it flows downstream. Increased biodiversity in the fields also promotes natural pest control and pollination, and conserves threatened species.[16]

Some of the most impressive evidence for the feasibility of these strategies comes from two field experiments conducted at Iowa State University's research farm in Boone County. First, Adam Davis and his colleagues tested whether increasing the diversity of crops grown in a single area could improve yield and profitability with fewer fertilizers and pesticides. The authors varied the numbers of crop species planted in rotation as well as the application of herbicides and fertilizer. They found that increased crop diversification maintained production and profitability while dramatically reducing expensive and harmful chemical applications. Longer rotations, whereby one or two different species are grown and harvested between each round of corn and soy, increased corn and soy yields from 4 to 9 percent over the traditional two-year rotation. Profitability was similar across all three treatments, but nitrogen fertilizer and herbicide inputs were reduced by as much as 80 to almost 90 percent. Amazingly, freshwater toxicity for the two longer rotations fell to a level two hundred times below that which occurred when conventional agricultural methods were used.[17]

In the second experiment, Meghann Jarchow and her collaborators evaluated tradeoffs between the environmental costs and economic benefits of bioenergy production from various combinations of corn, soy, and rye compared with newly planted native prairie vegetation. The prairie ecosystem that formerly dominated the Heartland has been virtually eliminated by the industrial farming of corn and soy. But prairie vegetation has many attractive properties when compared with corn and soybeans. It concentrates more organic carbon in the soils, improves soil stability, increases the complexity of the food web, and lowers the amount of nitrogen entering groundwater, lakes, and streams. In a companion study, Matt Liebman and Lisa Schulte showed that reincorporating prairie vegetation into agricultural landscapes greatly reduced water and sediment runoff from fields, and increased plant biodiversity by a whopping 380 percent.[18]

Jarchow found that total above-ground biomass production for the continuous corn plots was initially more than double that of the unfertilized prairie plots, with the fertilized prairie treatment coming in at halfway between. As the prairie systems matured over time, however, the initial differences were reduced more than in previous studies. As a natural perennial ecosystem, the prairie will almost certainly increase its productivity as the plants reach full maturity. Moreover, the root biomass was nine times greater for unfertilized prairie perennials than for either the continuous corn or the corn-soybean rotation. Greater root biomass stabilizes soils and decreases their vulnerability to erosion during flood rains.[19]

Net energy balance—that is, the amount of energy that is achieved from a crop after the energy inputs are subtracted—was greater for traditional corn. But in terms of the costs and benefits of nitrogen use, the situation was reversed. The quantities and cost of nitrogen inputs, and the concomitant risks of nitrogen pollution and the costs of removing nitrogen that are such a feature of continuous corn treatments, were nil for unfertilized prairie. Jarchow and her colleagues concluded that, while the net bioenergy produced from unfertilized prairie grasslands is less than from corn, those differences decrease as prairie fields mature. Critically, too, there are no environmental risks from farming prairie grasses, and the process is sustainable for decades to come.[20]

Biomass production is just one benefit of reconstituted prairie agriculture. Entire fields or strategically situated buffer strips can minimize soil loss and nutrient runoff, maximize biodiversity, and restore natural pest control and pollination.[21] Let's consider the benefits of each in turn.

It has long been recognized that riparian buffers composed of natural vegetation can reduce runoff and nutrient loss, but questions remained about which plants are best to use and how wide these strips need to be in order to be effective. To address these questions, Paul Mayer analyzed the results of forty-five published studies of eighty-nine riparian buffers. The average effectiveness of nitrogen removal in these buffers varied from 42 to 85 percent, and depended primarily on the width of the buffer zone (up to about 165 feet) and secondarily on the nature of its vegetation. Mayer's results have been confirmed experimentally. Prairie filter strips effectively reduced surface nutrient runoff in all situations compared to sites without buffers. Overall, the strips reduced nitrate to nitrogen runoff by 67 percent, total nitrogen by 84 percent, and total phosphorus by a remarkable 90 percent, values consistent with Mayer's earlier analysis. Importantly, these benefits were greatest in those years when there were high levels of rainfall concentrated in exceptionally strong storms. These kinds of storms cause massive losses of soil and nutrients, are the events most destructive to industrial corn and soy production, and are increasing in severity and frequency with climate change. In the United States from 1980 to 2016, $180 billion, or more than $2 billion per storm, was lost from severe storms other than hurricanes.[22] While not all losses were agricultural, droughts and severe storms together cost Americans more than $400 billion nationwide, and much of that came out of the Heartland.

Prairie strips also lead to gains in biodiversity. Plant diversity in six reconstructed prairie strips was nearly four times higher than for cropland. Moreover, diversity increased during the five years of the study as the prairie strips matured, and plant cover increased to over 100 percent because of plants overgrowing other plants. Bird abundance also increased nearly threefold. Fifty-two bird species were observed in the strips, although just sixteen species comprised 99 percent of all observations.[23]

New work in Michigan and Wisconsin confirms the results in Iowa. Benefits to the ecosystem in perennial fields were spectacular, and further benefits to biodiversity and crucial ecosystem services such as pollination and pest control were added. In the perennial fields, the sheer numbers of species—as well as the number of species from higher taxonomic groups—of plants, soil bacteria that consume methane (a potent greenhouse gas), plant-eating insects, predatory and parasitic insects, bees, and breeding birds were consistently greater than in the cornfields. Although the aboveground biomass production was more than three times greater for corn than for perennials, once again perennial fields were consistently superior in providing benefits to the ecosystem. Bacterial consumption of methane was several-fold greater. Sunflower pollination improved. Insects consumed more eggs of pests. Aphids declined and grassland bird species were more numerous.[24]

The overwhelming and sustainable benefits of this regenerative and multispecies agriculture for food, biofuel production, the environment, and human health mean that it should be adopted throughout the Heartland. Rotating a diverse array of crops; planting prairie strips to control surface runoff of nutrients, pesticides, and soil; and producing biofuels from perennial crops are all strategies that together can increase U.S. agriculture's resilience to climate change and maintain biodiversity throughout the Heartland.

But the thousands of miles of drainage tiles beneath Midwestern fields still discharge directly into ditches, streams, and rivers, contaminating drinking water and fueling dead zones in the Great Lakes and the Gulf of Mexico. Midwestern fields are routinely overfertilized by at least 35 percent, and the excess nitrogen and phosphorus escape to watersheds through these tiles. Reducing this pollution will require reducing fertilizer applications to the absolute minimum needed for optimal production. It will also require installing catchments and technologies such as woodchip bioreactors to remove most of the nutrients before the water escapes downstream. Recommendations for the frugal application of fertilizer and the imposition of a tax on its use could strengthen conservation efforts at the source.[25]

So could removing subsidies. Government programs subsidize bio-fuel production from corn by requiring that gasoline for cars include 10 percent ethanol. That requirement shores up the bottom line against direct farm expenses of genetically modified seeds, fertilizer, pesticides, and machinery. But the bottom line doesn't include the accumulating costs of cleaning up the contaminated drinking water in the Heartland and the growing dead zones in the Great Lakes and the Gulf of Mexico. Ethanol production as it is today would not exist in a free market, something that we as a nation supposedly believe in. As Amory Lovins demonstrated in *Reinventing Fire*, we could make biofuel from many other sources. Besides harvesting native perennials, we could recapture agricultural, forestry, and mill residues and utilize municipal solid wastes. These alternative strategies would sidestep many of the deep ethical issues that are raised when land that formerly produced food goes to making biofuels out of corn.[26]

We still need biofuels; we just need to produce them more responsibly. Even with electric cars and power generated renewably, heavy trucks, airplanes, and construction equipment cannot run on electricity alone—not yet anyway—and generators and machinery in remote areas still require liquid fuels. Nevertheless, we can make planes and trucks 50 percent more efficient right now. We can also cut down on fuel-guzzling commuter flights between regional hubs by building more energy-efficient and convenient passenger rail connections between cities, like Japan and much of Europe have done. In contrast, Amtrak trains in 2017 are only slightly more energy-efficient per passenger mile than planes. But for all their faults, U.S. trains are increasingly popular. Amtrak trains jolting along the Northeast corridor on third-world tracks still get you from downtown Washington, D.C., to midtown Manhattan faster and easier than getting to the airport, dealing with security checks and boarding, going up, going down, waiting for a gate, and taking the long taxi ride in impossible traffic from La Guardia to Midtown. Trains are packed and the energy savings would be enormous if the system were modernized.[27]

Moving east once again, we find agriculture in the Northeast poised for a renaissance. About 57 million people—roughly 18 percent of the U.S. population—live in the Northeast, including New England, New York, New Jersey, and Pennsylvania. The region generates $3.75 trillion in economic output, meaning that if the Northeast were an independent country it would have the fourth largest economy in the world. People increasingly want to buy higher-quality and local produce and have the money to pay for it. Forest cover in New England is more than twice the national average, at about 60 to 80 percent, with 12 percent of the land in crops, and 4 percent as grassland or range. There is a growing conservation ethic and interest in local produce. The Northeast is also by far the most urban region in the United States.[28]

Northeastern agriculture is a highly diverse and significant industry that comprises more than $71 billion in output and over 400,000 jobs. The most important sectors are dairy cattle and milk production, with $4 billion in output, followed by greenhouse production and nurseries. Top industrial processing sectors include paper mills (trees are grown for pulp), fluid milk processing, and fruit and vegetable canning, pickling, and drying, with a combined output of more than $27 billion. An increasing proportion of vegetable and fruit production is organic or pesticide-free.[29]

Farmers in the Northeast, like farmers everywhere, have to contend with climate change. Yet in doing so, these farmers are finding benefits as well as costs. Longer, warmer, and frost-free growing seasons may greatly enhance production of highly desirable and lucrative crops in the region. Bolstered by greenhouse production, Maine has become a major supplier of winter tomatoes. Traditional varieties of cultivated fruits such as raspberries and blueberries can be replaced by similar ones grown farther south. Ultimately, Northeastern agriculture may thrive under warmer conditions because freshwater

is plentiful, particularly compared with California and the Heartland. Improved water management with new reservoirs to store runoff from torrential rains may supply adequate water for irrigation
even in periods of drought. Crops requiring longer growing seasons—
including tomatoes, peppers, peaches, red wine grapes, and even
melons—are already profitable for some farms, and hold great potential for expanded production throughout the region.[30]

Organic food production is booming, but it cannot keep up with
the exploding demand. Most organic food is grown along the Pacific
coast, in the Upper Midwest, and in the Northeast, which has the
highest proportion of organic producers. Vermont and Maine lead
the pack with 5 and 6 percent of their farmers certified as organic
producers, respectively. (Many more farm organically, but forgo expensive USDA certification.) The market share for organic products
nationwide in 2014 ranged from about 6 to 7 percent for fruits, vegetables, and eggs to 14 percent for milk. Overall, the organic share
has shown double-digit growth most years since 2000 with no signs
of slowing down. Nearly half of all Americans polled said that they
actively try to include organic food in their diets, a trend that is especially strong among millennials. By 2016, growth in organic foods was
the top trend in the American food industry and 81 percent of American families were buying organic foods at least some of the time.[31]

Profits are also greater for organic farming than for conventional
farming. The Rodale Institute has conducted a thirty-year experiment comparing side by side the yields, profits, energy inputs, and
greenhouse gas emissions of organic versus conventional farming of
corn and soy.[32] Organic yields matched those from conventional
farming and significantly outperformed conventional farming in
drought years. Organic farming also helped to build rather than de-
plete soil organic matter, used 40 percent less energy, and released
only two-thirds the amount of greenhouse gases that conventional
farming does to obtain comparable yields. All this while turning
three times the profit of conventional agriculture.

The trend toward organic farming is obviously excellent news for
environmental sustainability and human health. But it is important

to remember that regenerative agriculture is not necessarily organic, and that most, if not all, of the environmental and societal benefits can be achieved even when it is not. Ecological (regenerative) farming is clearly the way of the future. Ecological farming focuses on redesigning whole farming systems as a human-managed ecosystem through a wide variety of complementary practices. These include reduced or zero tillage, more complex crop rotation, use of cover crops to build fertility and protect soils, reliance on soil biology rather than fertilizers for soil structure and fertility, use of composts, minimal use of agrochemicals, biological pest control, planting mixed crops, working with livestock using strategies like pasture and agroforestry, and landscaping for water use and conservation. In other words, ecological farming steers clear of unnecessary inputs, harnesses natural processes, and avoids doing harm to the local environment, to biodiversity, and to the fundamental biogeochemical cycles on which all life depends.[33] What could be better than that?

The agro-industrial giants and their sycophants and representatives have done an amazing job of extolling the virtues of industrial agriculture and dismissing ecological farming as a tree-hugging, hippy enterprise that could never feed the starving hordes. Generally, the U.S. Department of Agriculture and Congress have been happy to support them. But as this chapter documents, and as an increasing number of very smart businesspeople and investors are discovering, ecological farming can be brought to scale as the predominant strategy for feeding the world.[34]

EPILOGUE
AMERICA AT THE CROSSROADS

We are like tenant farmers chopping down the fence around our
house for fuel when we should be using Nature's inexhaustible
sources of energy—sun, wind and tide . . . I'd put my money on the
sun and solar energy. What a source of power! I hope we don't
have to wait until oil and coal run out before we tackle that.

—Thomas Edison to Henry Ford, 1931

America faces a great environmental crisis that threatens our
economy, security, and health. But it's an unnecessary crisis. We
know how to take steps right now that will minimize risks and improve our lives—and in many places, we're already taking them.

Severe, perhaps permanent, drought punctuated by extreme storms
and floods threatens the Southwest. But a combination of improved
water conservation, new reservoirs and irrigation systems that minimize evaporation, and more rational agriculture can meet most of
the region's water needs. In the Heartland and downstream, in the
bayous of Louisiana and the Gulf of Mexico, extreme weather events,
Roundup-resistant superweeds, soil erosion, nutrient overload, and
toxic drinking water threaten residents' livelihoods and wellbeing.
But adoption of ecological farming strategies—shifting to perennials for biofuel production, incorporating prairie buffer strips and
catchments to reduce nutrient pollution, and planting diversified
mixtures of crops—could sustain the region's agricultural productivity and profits while minimizing environmental damage and improving human health. And on the East Coast and around the Gulf, a
rise in sea level poses a grave and imminent threat. But a balanced
combination of mitigation and adaptation could stave off the worst-
case scenarios of retreat and resettlement, at least for all but the
most intractable locations. Massive investments in wind, solar, and

other forms of renewable energy are already easing future climate change scenarios while improving air and water quality today. All of these approaches would strengthen America's decaying infrastructure, create jobs, and promote sustainable economic growth.

Almost everybody knows this. How could we not? You can't live in California or other Southwest desert states turned brown by withered grass without experiencing the new reality of water rationing. Flood rains in the winter of 2016–2017 provided a welcome temporary respite in California. But without enough dams and reservoirs to save the excess water, most of it flowed out to sea. It will take a lot more rainy years to replenish groundwaters depleted by decades of excessive use. You can't live in the Midwest without hearing about toxic drinking water in Toledo, Ohio, and along western Lake Erie. You can't live on the East or Gulf coasts without fearing hurricanes and floods—especially in places like Norfolk and Miami Beach that are flooded regularly by full-moon tides. Memories of Katrina and Sandy have been burned into people's brains, and now there's Harvey, Irma, and Maria all in a single year with mounting worries that 2017 could be an inkling of the new normal. This has never happened before, contrary to President Trump's assertions. Among rational and informed people, however, the debate is over: costly and dangerous extreme weather events are increasing in frequency and intensity because of climate change.

Even the most rabid climate change deniers in the U.S. Congress almost certainly know that climate change is real. But they're trapped by reelection cycles and fear the massive amounts of dark money that would be pitted against them if they dared to cross the energy and agricultural lobbies. This was exactly how the tobacco industry waged a decades-long war of disinformation to conceal the catastrophic effects of tobacco on public health. Who doesn't know someone who died of smoking? And it is increasingly evident that big oil has waged the same kind of war to discredit basic high-school science: it is clear that burning fossil fuels really does release greenhouse gases that really do cause global warning. This outcome was even predicted in 1896 by Svante Arrhenius, who won the Nobel Prize in 1903 for his brilliance in physics and chemistry. His basic equation defining the relationship between carbon dioxide and temperature is still in use today. No one ever questioned the result until the big energy companies realized they had a problem and manufactured a false debate. Likewise, pesticide and fertilizer companies deceptively downplay the harmful consequences of industrialized agriculture. Upton Sinclair, who exposed corporate corruption a century ago, observed: "It is difficult to get a man to understand something, when his salary depends on his not understanding it."[1]

Nevertheless, environmental progress is ongoing and its momentum is increasing. The newly empowered Republican Right is doing its best to erase as fast as possible what has been accomplished so far. President Trump has also invited international rebuke by pulling out of the Paris climate accords that has been ratified by every other nation in the world—an action strongly opposed by the majority of American citizens and the major Fortune 500 American corporations, including ExxonMobil, whose CEO wrote personally to the president asking him not to withdraw. Moreover, in the face of continued Republican intransigence, more and more states, cities, and businesses are going it alone to adopt forward-thinking policies to reduce greenhouse gas emissions by investing heavily in renewable energy and increased efficiencies. The leading states in this movement—mostly blue—constitute two-thirds of the entire U.S. economy. Foreign states are now bypassing Washington to deal with these states directly.[2]

Embracing Renewable Energy

Energy consumption in Western societies has been excessively wasteful—which means that simply implementing existing technologies more fully can lead to enormous and immediate savings in both energy use and money spent. This basic insight comes from three major research efforts. The first is the analysis by Steve Pacala and Robert Socolow of "stabilization wedges" to eliminate the growth of carbon emissions—an approach recently explored in fascinating detail in Paul Hawken's popular and accessible book *Drawdown*, which is a virtual encyclopedia of practical solutions for combating climate change (complete with photographs and estimated economic costs and benefits). The second includes Amory Lovins's efforts to help private businesses save money and reduce climate change by reducing waste and adopting renewables. The third is the development by Mark Jacobson and colleagues of roadmaps for all fifty states to meet 100 percent of their energy needs with clean and renewable sources by 2050.[3]

Pacala and Socolow identified fifteen options for decreasing global carbon dioxide emissions over fifty years. They grouped these options into three categories: efficiency and conservation; decarbonization of electricity and fuels; and biological sequestration of carbon dioxide through improved management of forests and soils. Consider improved fuel economy. "Suppose that in 2054, 2 billion cars (roughly four times as many as today) average 10,000 miles per year (as they do today). One wedge would be achieved if, instead of averaging 30 miles per gallon (mpg) on conventional fuel, cars in 2054 averaged 60 mpg, with fuel type and distance traveled unchanged. . . . A wedge would also be achieved if the average fuel economy of the 2 billion 2054 cars were 30 mpg, but the annual distance traveled were 5,000 miles instead of 10,000 miles." Pacala and Socolow's recommendations fell on deaf ears during the George W. Bush administration, which insisted that the tools were not yet available to address climate change. But the common sense behind their proposals percolated widely through scientific circles.[4]

Amory Lovins also demonstrated how increased efficiency combined with renewable sources could replace the energy we currently derive from nuclear power and fossil fuels. But he did so as a consultant to major global businesses. Lovins is a physicist and former don of Oxford (the youngest ever, at age twenty-one). His book *Soft Energy Paths* showed how simple, but rigorously implemented, conservation measures from improved light bulbs to redesigned buildings would save untold amounts of energy, carbon dioxide emissions and, importantly, lots of money for the smartest American corporations. Basically companies could increase their profits manyfold by dropping archaic ideologies and hewing to the bottom line. They could do good by doing right. His message resonated with corporate executives at Carpet One, 3M Corporation, Exelon, Google, and other Silicon Valley businesses, which began to save billions of stockholders' dollars by implementing soft-path efficiencies.[5]

Lovins's 2011 book, *Reinventing Fire*, summed up his sermon to the private sector, arguing that big efforts to reduce climate change had already begun and that the financial savings were huge.[6] He made four major predictions. The first is that progress on reducing climate change will be led more by individual countries and corporations than by international treaties and organizations. The second is that the private sector and communities will play a greater role in reducing climate change than governments. The third is that developing countries will lead more than developed countries, which are saddled with decaying infrastructure. The fourth is that advances in clean energy and efficiency will be more effective than any future carbon pricing. Big changes have begun and the financial savings have been huge. If smart old money was in oil and coal, smart new money is in renewable sources of electricity.

Lovins revisited his predictions in "Four Trends Driving Profitable Climate Protection," an article that appeared in *Forbes* at the start of Climate Week NYC 2016. Its upbeat tone and the remarkable progress it records are a refreshing antidote to the consistent doom and gloom that characterize most of the dialogue about climate change. The numbers are striking. In 2013 alone, $650 billion was invested in renewables (not including hydropower), increased energy efficiency, and cogeneration of electricity with useful heat. Absent federal subsidies, wind and solar power still average less than 4 and 6 cents per kilowatt-hour, respectively. That is one-half to two-thirds cheaper than the energy produced by new fossil fuel plants, despite enormous federal subsidies to the oil and gas industries.[7]

Why do these pernicious subsidies exist? Unrelenting lobbying by oil and gas companies has distorted public discussion for decades. Their contributions to members of Congress have more than doubled since 2010: about $100 million was donated in 2016 with a party split of 7 to 1 in favor of Republicans. The majority leader of the Senate, speaker of the House, senators from Texas, and chair of the Environment and Public Works Committee receive the greatest contributions. It seems like more than a coincidence that all these federal leaders, despite the enormous weight of scientific evidence, consistently describe global warming as a scientific hoax.[8]

Given all the new investment, renewables are expected to double their capacity over the next fifteen years, while growth of fossil fuel plants will decline proportionately. Already, in both 2015 and 2016, global investment in renewables exceeded investments in fossil fuels two-fold. Even more remarkable, increased efficiency dwarfs the gains from increased renewables. To put this in perspective, energy savings in the United States due primarily to smarter technologies "have cut cumulative energy use *31 times more* than renewable growth raised supplies."[9]

Developed countries are using less energy and relying increasingly on renewables. The percentage of electricity generated from renewables in 2016 (excluding large hydropower) ranged from lows of 9 to 12 percent in the United States, Canada, Australia, and Japan to highs of 25 to 30 percent in the United Kingdom, Italy, Germany, and Spain and more than 40 percent in Denmark. Similar trends are evident in the developing world, notably in China, which is aggressively reducing its use of coal and investing heavily in solar and wind power. In 2013 China added more solar capacity than the United States has installed in the previous sixty-one years—a particularly shameful statistic since the technology was first developed in the United States. China generated 10 percent of its electricity from renewables in 2016, much like in the United States.[10]

Similar trends are occurring in India, South America, the Caribbean, and northern Africa. With few exceptions other than China, most of these advances have been achieved through private and corporate investment rather than government actions, which tend to be erratic or dysfunctional. The Paris Agreement to limit emissions, ratified by 194 countries by December 2016, is a magnificent international achievement that provides binding guidelines for the future as well as a vitally important sense of optimism. Nevertheless, all the achievements described here have been accomplished through economic common sense and a pragmatic corporate and social commitment to act in the common interest. In Lovins's words: "The transformation to a low-carbon future is arguably the greatest business opportunity of our times."[11]

Lovins's focus is on global patterns and trends. In contrast, Mark Jacobson and his colleagues conducted a detailed analysis of energy usage in each of the fifty U.S. states in order to construct state-by-state roadmaps for achieving complete carbon neutrality by 2050. Ponder that carefully. Their analysis shows how all fifty states could generate all of their power from renewable sources, eliminating the need for fossil fuels and nuclear power plants. The roadmaps would also generate new jobs, stabilize energy prices, and prevent the premature deaths of the approximately 63,000 Americans who die from air-pollution-related illnesses every year.[12]

The team investigated energy use in four sectors: residential, transportation, industrial, and commercial. First they determined the state's current sources and amounts of fuel consumed, including oil, gas, coal, nuclear, and renewables. Then they calculated what steps could be taken to generate all power needs within the state from renewables, chiefly wind and solar but also hydropower and other sources—always keeping in mind the alternative strengths of each state, so that the approach would be local and politically attractive. States like Texas, for instance, would build on their already enormous wind power capabilities. Texas currently gets 8 percent of its power from wind, which is three times more than any other state in total volume. Indeed, Texas ranks sixth in wind power generated and consumed among countries of the world—after China, the rest of the United States in toto, Germany, Spain, and India. Deeply conservative Georgetown, Texas, is on the verge of receiving 100 percent of its energy from wind power and a smattering of other renewables.[13]

Jacobson is a highly regarded energy-systems engineer from Stanford, but his analysis and conclusions have been criticized savagely by a host of equally distinguished scientists and engineers in the same *Proceedings of the National Academy of Sciences* where Jacobson published his paper. Jacobson has fought back hard and the debate has become unseemly. It seems to us that the truth lies somewhere in between the two points of view. The critics powerfully and convincingly argue that it is naïve to expect full reliance on renewables so soon, if indeed ever. But at the same time, progress in development of renewables is genuinely exploding worldwide—not least because, as Lovins has powerfully and repeatedly demonstrated, the smart money is being drawn to renewables because it makes the best sense economically. Indeed, changes are happening so fast that we can't help but be suspicious when people say that rapid transformation is impossible. It's possible, for example, that the problem of intermittent energy production by wind and solar in any one place could be overcome by improvements in the transmission infrastructure of the entire U.S. national grid by taking advantage of the fact that the sun shines or the wind blows somewhere in the continental United States virtually all the time, thereby eliminating the need for massive battery storage.[14]

The transformation to renewables may not be complete, but that's almost not the point. That it's happening very rapidly is undeniable. And it's not just the electric power industry. Who would have thought even a few years ago that a major automobile company like Volvo would commit all of its resources to manufacture only electric or hybrid cars by 2019? That's an enormous risk that this established, successful company would never have made if it didn't see that the handwriting on the wall is for renewables. And it's not just Volvo and Tesla. All of the major carmakers are beating the same drum to some degree.[15]

Indeed, in 2016 nine U.S. states produced more than a fifth of the electricity that they generated in-state from wind and solar sources: Vermont, 100 percent; Maine, 54 percent; Idaho, 47 percent; Iowa, 39 percent; Kansas, South Dakota, and California 30 percent each; Oklahoma, 26 percent; and Minnesota, 23 percent.[16] Ironically, only one of these states, California, is solidly Democrat. Most are solidly Republican. For these Republicans, the decision to adopt renewables is strictly business. Lovins would be proud. Western states have loads of wind power and have decided to take advantage of it. The consequences for increased employment are also great. One of the fastest-growing, highest-paid jobs in the United States is wind power technician.[17] Ironically, nothing appears to have changed politically; members of Congress from these same Republican states still support subsidies to the oil and gas industry and oppose restrictions on greenhouse gas emissions. Nevertheless, these states' use of renewable energy is helping to slow the rate of global climate change.

Many cities, and a majority of citizens, have also begun to opt for climate mitigation. One of the most striking examples is San Diego, the eighth largest U.S. city. Led by Republican mayor Kevin Faulconer and a Democratic city council, San Diego in 2015 signed a legally binding plan to run on 100 percent renewable energy by 2035, the largest American city yet to do so. The plan also calls for reducing the city's greenhouse gas emissions by 80 percent by 2050. This remarkable bipartisan measure was featured in the *Guardian* as an example of how the destructive American debates about climate change could have been avoided through a nonideological examination of local issues and common sense. Faulconer was a volunteer for San Diego parks before becoming mayor. His commitment to the environment is strong and he used his influence with the Republican establishment to convince them that the plan was smart economics as well as the right thing to do for the future of the city. "This isn't a partisan issue," Faulconer told the *Guardian*. "I pride myself on being fiscally responsible and environmentally conscious. The two aren't exclusive. I've never seen it as a zero-sum game."[18]

The business community bought into a vision of clean-tech jobs, low-carbon innovation, and clean air, knowing that San Diego is already a major tourist destination. The city gets 40 percent of its power from clean energy, and already ranks second in the United States in electrical output from solar energy. Bringing in entrepreneurial clean-tech companies fits the "Smart City" strategy for a region that is already a world center for wireless, biotech, medical devices, genetics, big data, and advanced defense. In accordance with the plan, by 2050 half the city's vehicle fleet will convert to electric, and virtually all methane emissions from water treatment and sewage will be captured and recycled. The city's tree canopy will be increased to 35 percent, to cool the town, clean the air, and soak up carbon.[19]

Steve witnessed an aspect of San Diego's vibrant science and technology culture from the flight deck of the *USS Midway*, the giant World War II aircraft carrier anchored as a museum at the city center pier. The occasion, a STEAM (science, technology, engineering, arts, and mathematics) event sponsored by the San Diego Unified School District, showcased electric vehicles and charging machines before hundreds of cheering high-school students. The city utility, San Diego Gas and Electric, a partner of the plan, has installed 3,500 electric vehicle chargers throughout residential neighborhoods to encourage electric vehicle buy-in in a city that already has the highest per capita ownership of Teslas, and 21,000 electric cars overall.[20]

The chief sustainability officer of San Diego is Cindy ("Cody") Hooven, thirty-nine, a graduate of the Center for Marine Biodiversity and Conservation at the Scripps Institution of Oceanography. Steve met with Hooven at a "sustainability hackathon" on the University of California San Diego campus that further exemplified the changing culture of the city. For the event, 120 young engineers and IT types, students, and mentors were engaged in a seventy-two-hour marathon competition to generate ideas to help solve San Diego's climate sustainability issues. Components of climate resiliency considered were energy- and water-efficient buildings; gas and waste management for zero net waste; clean and renewable energy; land use; and transportation including walking, cycling, and mass transit. The teams would be judged on the practicality of their creative solutions, with the winning team receiving $5,000 along with other prizes. Sponsors included Qualcomm, the wireless chip company Teradata, and an assortment of clean-tech companies on the lookout for new talent.[21]

Cody told Steve that engineering schools in San Diego are booming. The city wants to exploit its talent and new business models to create more jobs for solutions-oriented people. While they were talking, the two were joined by Daniel Obodovski, lead organizer and co-author of *The Silent Intelligence*. Obodovski is an expert in what is called the "Internet of things," the vast, evolving wireless system that connects thermostats, appliances, self-driving cars, and much more to smartphones and laptops so they can be controlled from wherever. He explained that a hackathon is a tool to better understand the critical role of technology to measure things like water use or temperatures in buildings, and then use the data to create credible milestones. The idea is to use new sensors, cull big data, and engineer feedback loops to increase efficiency and create more jobs and revenues. "Hopefully, the people in the room will commercialize these ideas with a sponsor in an incubator or a startup. The ultimate goal is to confront climate change. We are saving [the planet] not only for ourselves, but for our children."[22]

Hooven described the San Diego plan's vision of mixed communities that include both jobs and homes in close proximity, a strategy that would shorten today's lengthy commutes. The two biggest goals of the plan are energy reduction and smarter transportation, with half of the city's transportation to be shifted to walking, biking, and public transport by 2025. These measures will be supported by a half-cent sales tax and the city will take more control of energy production, including encouragement of rooftop solar panels and solar farms in the neighboring deserts east of San Diego and in Baja California. The city is also addressing sea level rise. Mission Bay already floods and eelgrass comes up storm drains in the middle of airport runways.[23]

Hooven is upbeat. "Everybody's taking the future seriously in San Diego and, like southern Florida, we have formed a regional climate cooperative with other cities, and the San Diego Foundation, the airport, the port, and Sacramento listen to us. For drought, we are investing millions in what we call Pure Water—"

"You mean toilet-to-tap?" Steve asks.

Hooven laughs. "Pure Water sounds more delicious. But we are too dependent on the Colorado and the Sacramento [rivers], and those ducts would be shut down in an earthquake, so there is a resilience side to climate change [adaptation,] as well as the [benefit of] CO_2 [being] reduced from not relying on the huge pumps."[24]

Seeking Agricultural Sustainability

American agriculture is unsustainable because it relies on methods of production that cause acute environmental damage and threaten human health both directly through poisoned drinking water and indirectly by the promotion of unhealthy diets. It is also unsustainable because we waste roughly half of all the food we produce because food is cheap, Americans are exceptionally picky, supply chains are inefficient, and supermarkets, food processors, restaurants, and individuals at home discard huge amounts. Wasted food is the major component of U.S. landfills.[25]

The problem of where to put the waste has gotten so bad that major cities are moving to require that organic waste be recycled as biomass for energy and compost. New York City already has the largest residential program in the country, but it still accommodates only a small fraction of the millions of tons of organic waste thrown away each year. Full participation by the 24,000 large residential buildings in the city (with more than eleven apartments each) will require entirely new recycling systems to accommodate all that organic waste. Wasting food is a global scandal that demands major shifts in attitude to resolve. It is also a climate change problem because wasted food in fields and landfills is a major source of greenhouse gas emissions.[26]

The major agricultural giants like Archer Daniels Midland, Monsanto, and DuPont have reaped enormous profits from subsidized biofuels and from GMO crops that are addicted to the panoply of poisons and fertilizers dumped on fields. The catastrophic losses of rich soil and pure drinking water that result from these practices should be considered collateral damage to the nation's natural capital.[27] Progress at the national level is unlikely while the Republicans control the presidency and Congress and while they continue to strip away environmental regulations. Nevertheless, we already know how to farm in highly productive ways without destroying the environment and while working to protect public health. Fortunately, more and more farmers are moving in that direction in spite of federal intransigence.

Progress toward sustainability may be just around the corner because Midwestern profits are falling fast. The ethanol boondoggle is in danger of collapsing due to overproduction and falling fuel prices. Farmers are being squeezed between their climbing costs for seeds, fertilizer, and equipment and grain prices that have been in free fall since 2014. The U.S. Department of Agriculture forecast for 2017 has real farm cash receipts down 14 percent from 2015 and down 36 percent from the previous high set in 2012, just after we first visited John Weber in Iowa. Farmers who borrowed to increase production when corn was through the roof are deeply in debt, with bankruptcies looming for many. International competition is also increasing. The American share of the global grain trade has fallen from 65 percent in the mid-1970s to 30 percent today.[28]

One sensible way out of this financial bind is to farm more sustainably and cheaply, and there are many ways to do that. The simplest, but largely overlooked, first step is to farm only those portions of land that can produce a profit. Not all "good Iowa dirt" is so good after all. Elke Brandes and her colleagues analyzed corn and soy production at a fine geographic scale for all of Iowa during the highly profitable years from 2010 to 2013, then used those data to predict profitability plot by plot for 2015.[29] Their forecast? More than 6 million acres, one quarter of all Iowa farmland, would lose more than a hundred dollars per acre due to the heterogeneity of the soils and other unfavorable environmental field conditions. Highly automated corn and soy farming over fields spanning many hundreds to thousands of acres can easily miss that some areas are failing to produce an adequate harvest. The scientists urge farmers to take marginal lands out of cultivation and convert them to perennial grasslands that can be harvested for biomass even as they reduce soil erosion and promote biodiversity.

A second, more proactive approach is to adopt integrated farm management strategies that combine traditional techniques of crop rotation with minimal use of modern chemicals. This approach builds on the results of crop diversity studies by Adam Davis and his colleagues described in Chapter 11.[30] It involves rotating a greater variety of crops than just traditional corn and soy, and incorporating livestock for their manure fertilizer. Pesticides and herbicides are employed only as needed. Herbicide use in the experiment was eight times less than for conventional corn and soy, and fertilizer use was down 86 percent. Those results translate into enormous savings for farmers operating on the edge, with the added benefit that virtually no toxic materials poison the surrounding waters. Importantly, crop biomass production in this study was as high as for conventional farming and the methods can be successfully employed on a commercial scale. More work is involved. Farmers need to be on top of things day to day to apply pesticides only as needed. But more work means more jobs, which is good for rural communities. The point is that ecological farming can compete successfully with industrial farming if farmers are willing to change.

A third alternative is to switch to perennial grains not only for biomass but also potentially for food production. We have already seen how biomass production from diverse prairie vegetation on marginal lands rivals or exceeds biomass production from corn—but without the expensive pesticides, fertilizer, water, superweeds, and pollution generated by industrial agriculture. The deep root systems of perennials help stabilize the soil and prevent the massive erosion that is increasingly common on Midwestern fields after extreme storms.

The challenge for food production, however, is to selectively breed perennial species to produce more abundant and larger edible grains. The greatest progress so far has been achieved with varieties of wild prairie wheatgrass developed at the Land Institute in Salina, Kansas.

Branded as Kernza, the grains are still considerably smaller than domesticated annual wheat, and bread baked from Kernza doesn't taste quite right to many people. Nevertheless, Kernza is beginning to reach a small niche market. It is also being used to produce whiskey and beer. Perennial wild sunflowers are also being bred to produce more and larger seeds that could be harvested for vegetable oil.[31]

Progress in perennial crop improvement is steady but slow. So far, scientists have mostly worked within the genomes of a single wild species to breed for favorable traits. Other options to speed things up include hybridization of perennial wheatgrass with more productive annual species and genetic engineering. Genetic engineering has a justifiably bad reputation among environmentalists due to the problems related to Roundup-ready GMO corn. But these problems are not the fault of genetic engineering per se but of what it was used to produce—namely, corn plants that cannot survive without massive amounts of pesticides, fertilizer, water, and farming methods that cause massive soil erosion. If GMO techniques were instead focused on inserting genes for greater grain size or productivity into perennial wild stocks, without sacrificing the inherent hardiness of the perennial species, that could be a very good thing.[32]

Finally, an increasingly profitable and popular approach is to abandon industrial grains entirely and switch to fruits and vegetables that offer higher returns on investment and do less environmental damage. Differences in profits can be striking. In a good year an acre of corn might yield $300 whereas the same acre of apples can make $2,000 or more. Ironically, Iowa and other Midwestern states were among the biggest producers of fruit and vegetable products until the 1950s, when two developments wiped out farmers of these crops. One was the rush to industrial farming of field corn and soy for export, animal feed, and eventually biofuels, a move promoted by massive federal subsidies. The other was crushing competition from high-quality, efficiently marketed Californian fruits and vegetables. But the effects of the California drought and the rise of new local and regional markets are fueling a resurgence in farming produce in the Midwest and east of the Mississippi generally. Specialty crops and niche markets offer exceptionally high returns, which should encourage local farmers to jump back into this market.[33]

Which brings us back to fruits and vegetables. An amazing 95 percent of the leafy greens grown in the United States still come from California and Arizona, with California producing 86 percent of leaf lettuce, 77 percent of romaine, 71 percent of iceberg, 66 percent of spinach, and 20 percent of cabbage. Percentages for other vegetables are just as amazing: California grows 99 percent of U.S. artichokes, 95 percent of celery, 89 percent of cauliflower, 69 percent of carrots . . . the list goes on and on. Most come from the so-called Salad Bowl of the World, located in the Salinas Basin on the Central Coast—a region that produces roughly 70 percent of all U.S. lettuce and provides 73,000 jobs and more than $5 billion in income to Monterey County. But Salinas farms also use 91 percent of the water drawn from a groundwater basin estimated to be over-drafted by some 150,000 to 300,000 acre-feet during drought years—which have become more and more common. So farmers have teamed up with the San Jose tech community to finetune their irrigation systems to use as little water as possible.[34]

There are still two obvious problems. The first is the uncertain future of California's water supply and the conflicting water needs of cities and suburbs versus agriculture. High-tech controls can reduce water consumption up to a point, but clearly there are limits. Given the nation's dependence on California vegetables, this is a food security issue.

The second, greater problem concerns the gross inefficiency and waste in getting produce from far western Salinas to American consumers farther east. More than half of the U.S. population lives east of the Mississippi River. Trucking California produce eastward consumes an enormous amount of fossil fuels and generates greenhouse gases that contribute to global warming. There are also questions of spoilage, risk of infection by *E. coli* and *Salmonella*, and loss of nutritional value. When the system works efficiently, it takes between fifty-six and eighty-two hours to get a head of lettuce from Salinas to New York City. The Natural Resources Defense Council estimates that getting food from the farm to the consumer nationwide uses 10 percent of the entire U.S. energy budget. The waste along the way is enormous: 24 percent of fruits and vegetables are lost in production, handling, and processing; another 12 percent are ruined in distribution and at retail stores; and 28 percent are lost as consumers store and use them in food services and households. Add it up, and two-thirds of all fruits and vegetables grown in the United States are never eaten.[35]

The time is ripe for entirely new approaches. One of the most promising is urban vertical farming, which produces the vegetables that people want practically next door.[36] The advantages are obvious if production can meet the great demand. Cropland and biodiversity could be restored to a more natural state while at the same time most of the harmful externalities of conventional agriculture would be eliminated. Savings in transportation costs alone could be enormous, and the food quality should be better and safer because the supply chain would be so short.

Vertical farming has been around for more than two decades, but is taking off only now, as major investors become involved. Vertical farms already exist in Houston, Seattle, Detroit, outside Chicago, and New York City. But the one getting the most attention is AeroFarms, which has built the largest vertical farm in the world in an abandoned factory in the heart of Newark, New Jersey. The aisles of the farm are tall. Layers upon layers of shelves support plants growing on a reusable cloth, illuminated by LED lights, and watered by aeroponic mist. Most tiers are reachable only by forklift crane. While the enterprise builds its market share, the farm will exclusively produce leafy greens that can bring premium prices. Later, after proof of concept, these vertical farms can be used to produce a far greater diversity of crops.

David Rosenberg is the CEO and cofounder of AeroFarms. Jeremy met Rosenberg in April 2017, when he chaired the session on green farming at the Smithsonian Earth Optimism Summit and described AeroFarm's achievements. The statistics are remarkable. On one acre of shelves, the farm produces greens equivalent to 130 acres of field production, using half the amount of fertilizer, just 5 percent of the water, and no pesticides, herbicides, or fungicides. Rosenberg has been strongly influenced by the basic principles laid out in William McDonough's classic book *Cradle to Cradle*. He sets AeroFarm's goals of eliminating waste, designing for disassembly, and using the right materials in the right places.[37]

To achieve these goals, AeroFarms employs a wide variety of experts. Biologists develop the physiological guidelines to optimize plant growth, nutrition, and taste. Engineers make it all work via the thousands of sensors that monitor environmental conditions on a micro scale. Data analysts monitor how well they are doing and tweak the system as needed. Conventional farmers face the challenges of varying weather conditions and pests, whereas AeroFarms has to figure out how to re-create what nature does in a field and make it better. The team investigates what light spectra and intensities to employ, which nutrients and micronutrients to add, and how to vary all of these things over time for the best possible outcomes.[38]

AeroFarms doesn't use GMO plants and is not involved in plant genetics. But its farmers and other staff are masters at controlling environmental conditions to nurture plants in different ways. A whole new science of fundamental plant biology is involved. Exploiting this knowledge has been key to their ability to conserve water. They are also developing environmentally sound ways to stress plants to grow differently, which allows them to manipulate taste and nutritional value. They can grow kale to be more tender or less bitter or amplify the pepperiness of arugula. So far, they have developed more than three hundred different varieties of leafy greens and, according to Rosenberg, chefs are enthusiastic about them.[39]

AeroFarms sells to major retailers like ShopRite and FreshDirect at the same price as conventional organic farmers charge, although its crops are not technically organic because the plants are not grown in soil. (Rosenberg describes their produce as "beyond organic.") This is a sensitive issue. Traditional organic farmers are outraged that similar systems have been ruled to be organic by the U.S.

Department of Agriculture. They object because no amount of gee-whiz chemistry and engineering can create the complexity of a natural soil with its panoply of bacteria, fungi, nematodes, and worms that undoubtedly affect plant growth. This doesn't detract from the ecological value of alternative indoor agriculture that is pesticide free, but it is not organic in the traditional sense.

AeroFarms isn't profitable yet, but it's getting there. Two of their major backers are Prudential Financial and Goldman Sachs. The whole enterprise strikes us as having the same feeling of excitement as in the early days of Amazon or Apple. The market is expected to quadruple to nearly $4 billion in the next few years. And that's just the beginning; the Northeast is full of abandoned mills and factories that could be the regional farms of the future.[40]

Responding to Sea Level Rise and Stronger Storms

Jeremy has been afraid of earthquakes ever since he saw the film *The City That Waits to Die* about San Francisco's vulnerability to earthquakes. So when he and his wife, Nancy, moved to the Scripps Institution of Oceanography, he got a geological map to figure out where they might want to live. There are beautiful houses high up on cliffs with astounding views of the ocean, but they mostly lie along the Rose Canyon Fault, which could slip dramatically during an earthquake. Another option was to live right on the shore, which is highly vulnerable to erosion—the terrain is unstable and walls and floors are known to crack as houses settle. In the end, they moved to a street in old La Jolla with no great views but on comparatively stable ground. Better safe than sorry.[41]

Unfortunately, most towns and cities aren't built with these considerations in mind. Typically the reasons for ignoring natural dangers are practical. Harbors and port facilities must be built on the shore. Transportation hubs and networks must be linked to port facilities. Building on flatter land is easier and cheaper. Boston, New York, Baltimore, Norfolk, Charleston, Savannah, Mobile, New Orleans, Galveston, San Diego, and San Francisco all owe their growth and economic success to their proximity to great harbors for commerce. Then there are the aesthetics of a coastal home. But that comes with the increasingly dangerous and extraordinarily costly threat of sea level rise fueled by climate change, accompanied by larger and stronger storms. Hurricane Sandy caused $65 billion in damages, of which just $30 billion were insured, and $19 billion in infrastructure damages. Nearly 90,000 buildings were flooded, 2 million people went without power for days, 1.1 million schoolchildren stayed home for a week, and forty-three people died. The subway system, the subterranean heart of the Big Apple, went into cardiac arrest, shutting down for the first time since 1904—that is, for the first time ever.[42]

Let's take Sandy and kick it up a notch. What would the costs of a rise in sea level be for the entire United States? Perhaps the best estimate comes from an analysis by James Neumann, Kerry Emanuel, and their colleagues at the end of 2014. (Emanuel, you may recall, is the MIT meteorologist who developed a Power Dissipation Index to explain the destructive potential of extreme storms in terms of their maximum sustained wind speed.) The Neumann team published their results in the journal *Climate Change* with the straightforward title "Joint Effects of Storm Surge and Sea-Level Rise on U.S. Coasts: New Economic Estimates of Impacts, Adaptation, and Benefits of Mitigation Policy."[43] The authors used models for tropical cyclones, storm surge, and economic impact and adaptation to examine seventeen multicounty areas along the East, Gulf, and West coasts. As expected, the results showed that large areas of coastal land and property are increasingly at risk as the sea level rises and storms intensify. They also demonstrated that adaptation is a cost-effective, if imperfect, response to risk. Using site-specific data, including local coastal geography and the slope of the bottom offshore, doubles the magnitude and costs of storm surge over those for sea level rise alone. This is especially true along the East and Gulf coasts with their broad and gradually shoaling continental shelves. Mitigating greenhouse gas emissions significantly would lower the expected losses. But the bottom line is close to $1 trillion in damages nationwide, with roughly half for relocation and another half for protective measures. (Loss of a large ice sheet in Greenland or Antarctica would bump up costs dramatically.) Remarkably, if that number seems exaggerated, Hurricanes Harvey, Irma, and Maria may have caused half of that amount of damage after all the chits come in, just in 2017. Already estimates for Harvey are reaching $180 billion, for Irma $83 billion, and for Maria $85 billion, for a total of $348 billion. Economists talk about how economies will bounce back fueled by all the new business for rebuilding. But that is very small consolation to the vast majority of homeowners and small businesses, only a small fraction of which have flood insurance.[44]

A trillion dollars is a very large amount of money and ought to be taken very seriously. We also need to expand our view beyond the quick fixes for the usual calamities and take into account that the sea level is rising every day. Measures to protect New York City from another Hurricane Sandy could easily fail ten years from now because the sea level will have risen just that teensy bit more—but a teensy bit that represents twice the global average rate of sea-level rise. Most folks haven't yet digested that nothing is forever. All the uncertainties about Greenland and Antarctica mean that no major coastal city can be considered safe within fifty to a hundred years. That doesn't mean that measures shouldn't be taken to protect the here and now; such measures must be put into effect as fast as possible. But measures also need to be taken on time scales much longer than the average political term in office or the average institutional funding cycle. Otherwise we will bounce along from one enormous crisis to the next. Europeans are pretty good at long-term planning: London has already retrofitted the Thames River Barrier to protect the city until 2100, and Hamburg has elevated parts of the city to withstand twenty-five-foot storm surges. But Americans are not.[45]

So let's examine again the situation for metropolitan New York City. Several excellent, popular articles appeared about this in 2016. Most highlight in one way or another Klaus Jacob's four complementary approaches that need to be incorporated in an overall plan. The first is installation of effective physical barriers to hold the water back.

Doing this right is incredibly expensive, but only a fraction of the cost of Sandy. The second is what the Dutch are doing in the Netherlands: living with and among rising waters, allowing them in so they can escape again quickly after severe flooding subsides. Jacob points out that the Dutch did exactly that when they first developed New Amsterdam in the seventeenth century. Water Street and Canal Street were what their names imply. Third, people need to start now to strategically relocate from the most vulnerable areas to higher ground. And finally, the risks and costs need to be spread among all of us through insurance.[46]

Plans are in the works and money is being committed for several measures to protect New York City from flooding. The biggest, most important, and most effective would be surge barriers that allow ship passage. The largest and most expensive barrier would extend from the Rockaways south of Jamaica Bay to Sandy Hook, New Jersey, to protect Lower New York Bay. The second, smaller barrier would be placed at the northern mouth of the East River somewhere east of LaGuardia Airport.[47] The Dutch have developed and installed these kinds of structures for decades, and they've proven effective. But at $20 billion or more, they are still too bitter a pill for the governor or mayor to swallow.

Even so, the city and state are moving forward with a number of ambitious plans to fortify the boundaries of the five boroughs against another storm. The most impressive is a U-shaped barrier of earthen berms, hardened walls, and gates that would keep floods out of Lower Manhattan, which is still the financial capital of America and the world. More than $600 million has already been obtained for this, more than a quarter of it from federal funding. But the project is still several hundred million dollars short and construction is several years from being finished. All entrances to the subway system will need to be elevated to keep water out. Elevated trains like the old Third Avenue El could also replace the most vulnerable parts of the system.[48]

Another ecologically motivated approach is to construct natural barriers all around the city, much as they existed in the past. Sixty

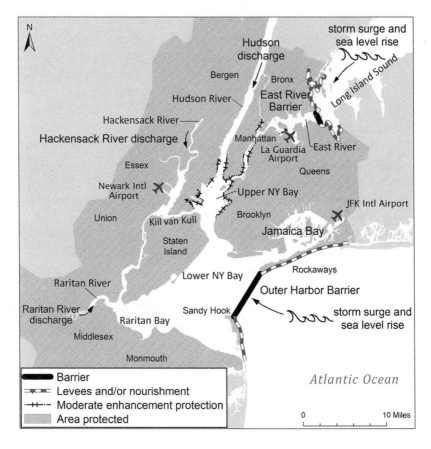

Strategy for the protection of the New York City metropolitan area, including adjacent New Jersey, based on the construction of two gigantic storm surge barriers. These barriers would extend across western Long Island Sound and outside Lower New York Bay to isolate the entire metropolitan region from storm surge. (The gray shading shows the protected area.) Adapted and redrawn from Jeroen C. J. H. Aerts et al., "Evaluating Flood Resilience Strategies for Coastal Megacities," *Science* 344 (2014): 473–475, doi: 10.1126/science.1248222. Reprinted with permission from AAAS.

Strategy for the protection of the New York City metropolitan area, including adjacent New Jersey, based on the construction of local barriers to protect the most vulnerable areas and critical infrastructure including subways, tunnels, and airports. Adapted and redrawn from Jeroen C. J. H. Aerts et al., "Evaluating Flood Resilience Strategies for Coastal Megacities," *Science* 344 (2014): 473–475, doi: 10.1126/science.1248222. Reprinted with permission from AAAS.

million dollars in federal funding is already in place to build living oyster reefs all along the south shore of Staten Island, and $7 million in state funding has been designated for onshore sand dunes, with beach grasses, recreational areas, and a breakwater for boat access. Much of the vulnerability of southern Manhattan is due to the fact that large portions of the financial district were constructed on top of wetlands and landfill. So the idea now is to rebuild more natural wetlands and keep developers out. The chronic problems of Co-op City in the Bronx are another cautionary example of what can happen when buildings are erected on landfill.[49]

New York City's plans are important, not least because they have a genuine chance of succeeding until the end of the century—a prospect that's virtually impossible for Miami or New Orleans, no matter what the Army Corps of Engineers may propose for the Big Easy. But New York is not the only place that is making serious strides. Government experts in Rhode Island, for example, are conducting the most objective coastal flooding evaluation analysis we have run across. They're doing it street by street and town by town, from one end of the state to the other. Rhode Island is the tiniest state but it has the second highest population density at a little over one thousand people per square mile. Its mean elevation is two hundred feet, roughly double that of Florida and Louisiana, but much of its coastal zone is very close to sea level and highly vulnerable to storms. People there have not forgotten the Great Hurricane of 1938, which was among the strongest and most destructive storms on record. Most of the hurricane's 682 deaths occurred in Rhode Island, and the entire state was devastated.[50]

Grover Fugate is executive director of the Rhode Island Coastal Resources Management Council and spearheads the development of the state's Shoreline Change Special Area Management Plan. He and Jeremy met briefly in Providence and he sent Jeremy detailed information about what his program is doing to alert Rhode Islanders to their high exposure to coastal flooding. The heart of the plan is a highly sophisticated, state-of-the-art online map resource called STORMTOOLS that displays coastal storm inundation with and without sea level rise for storms of differing intensity and return period. The resolution is extraordinary, six inches vertically and 3.3 feet horizontally, and it works for the entire coast. That's better resolution than the Federal Emergency Management Agency (FEMA) achieves.[51]

STORMTOOLS is free, easy to use, regularly updated, and accessible to anyone. A current homeowner or prospective buyer can type in an address and out pops the vulnerability of the property to flooding for different storm and sea level scenarios, allowing the individual to objectively assess the risk for that particular property. STORMTOOLS can also be used by coastal planners and for the engineering and design of buildings, neighborhoods, or entire towns. In other words, it provides exactly the kind of information that Jeremy was trying to pull together for buying a house in La Jolla, but for storm flooding rather than earthquakes.

Fugate's team used STORMTOOLS for a detailed assessment of the vulnerability of two Rhode Island towns that differ greatly in their shoreline situation and exposure. Despite their differences, estimates of the rise in storm water levels due to storm surge and tides are about fifteen feet at both locations. Estimates of the proportion of structures that will sustain 50 percent damage or more is about 20 percent in both cases. With six feet of sea level rise, the proportion of severely damaged structures would be nearly 100 percent. That's almost as bad as Miami, but at least Rhode Islanders can move to higher ground. Most importantly, the state is dedicated to providing local guidance to its citizens, so they know their options. Nobody is trying to hide the truth.[52]

Similarly important developments are suddenly happening on the Mississippi Delta through the efforts of a coalition of environmental groups and public citizens called RESTORE. An excellent website explains the coastal crisis, presents restoration solutions as defined by the newly approved Coastal Master Plan, and even provides an app for people to look up the vulnerability of their home to coastal flooding.[53] The coalition is effectively offering the same kinds of vitally important information that Fugate's group is providing for

Rhode Island. The difference is that it is a private effort, not a state government initiative, that is providing the truthful information that citizens need to plan their lives.

Public or private, nothing like these highly organized efforts to inform citizens in Rhode Island and Louisiana seems to be happening in Florida. Imagine the panic and collapse in the value of real estate that would occur if Florida engaged in the same kind of responsibly objective risk assessment for metropolitan Miami, Tampa, and St. Petersburg. In fact, the online real estate firm Zillow published in 2017 a nationwide survey of anticipated losses of homes by 2100 if the sea level rises by six feet—a number that we have seen is close to the emerging scientific consensus (assuming that nothing drastic happens soon in Greenland or Antarctica that would make the situation even worse). The results are sobering, especially because they don't take into account the poisoning of drinking water resources by saltwater intrusion into aquifers and the effects of storm surge over and above the actual sea level rise.[54]

The Zillow analysis concludes that more than three hundred U.S. cities would lose at least half their homes and that thirty-six such cities would be completely lost. One in eight Florida homes, or nearly one million homes in all, would be underwater—literally—which would mean overall losses of $413 billion. This figure would represent nearly half of the anticipated lost housing value of $882 billion nationwide.

Bloomberg Business News reported that some residents in south Florida already want to get out before any possible devaluation, but are having difficulty selling. Needless to say, the extensive damage and deaths caused by Irma, which did not even deliver a direct hit to major cities, will likely increase the potential exodus. The ironies are many. There is no oil or gas to speak of in Florida, yet the state's politicians and many of its residents deny the science behind rising seas. In effect they are in thrall to outside interests, to a fossil fuel industry that is slowly drowning their own state. Thus, the future of Florida's coasts is being determined by a few dry-land billionaires in Kansas and Texas, who, of course, fund their leaders. So the question is, at what point will Floridians choose between ideology and reality, between politics and real estate prices? Or will they be content to boil away slowly like frogs on a stove?

Recent research emphasizes that the consequences of global warming, drought, and sea level rise will be quite dire over the coming decades. Moreover, the effects will not be uniformly felt throughout the United States, and they will not be fair. Southern states will suffer far more than others, both in terms of economic losses and sweltering temperatures that will dry up fields and make urban life almost unbearable without air conditioning, which will be extremely costly. "Climate change will act as a kind of reverse Robin Hood," explained the *Atlantic*, "stealing from the poor and giving to the rich. It will impoverish many of the poorest communities in the country—arrayed across the south and southwest, and especially along the Gulf Coast—while increasing fortunes of cities and suburbs on both of the coasts."[55]

Calculations suggest that damages in the United States could reach 0.7 percent of total gross domestic product (GDP) for every one-degree Fahrenheit rise in global temperature. But the counties that will suffer most are in areas that are already hot, like Arizona and Texas, where losses could rise to 10 to 20 percent of local GDP if emissions were to continue to rise unchecked. Projections of heat-related deaths by 2100 would be on a par with the number of Americans killed annually in car crashes. If this sounds extreme, consider that more than 35,000 people died in 2003 during Europe's extreme heat wave.[56]

Maybe it's time to consider that move to Maine or Montana.

Many good and farsighted policies and actions are addressing the environmental challenges confronting us as a nation. Thus far, private industry, nonprofit organizations, and forward-thinking states, cities, and politicians are leading the charge. Meetings like the Smithsonian Earth Optimism Summit offer opportunities for those people and organizations making real and substantive progress to compare experiences and generate innovative ideas for moving forward.[57] But positive change would happen a lot faster and more efficiently if the federal government prioritized the future prosperity of its citizens over the demands of special interests.

The foot-dragging of the EPA with regard to dangerous pesticides and herbicides and to the poisoning of the Great Lakes and Midwestern groundwater cannot ethically be explained away as politics as usual. The Clean Air Act alone will have cumulatively saved 4.2 million lives and more than 230,000 lives every year by its fifty-seventh anniversary in 2020. The act has also saved $22 trillion in health care costs. People's lives are at stake every hour of every day when protections are rolled back or not enforced, as they have been in Flint, Michigan, or along Cancer Alley on the Mississippi Delta.[58] In addition to routine exposures to toxins, the recent flooding and explosion of chemical plants that produce toxic substances in Houston due to Hurricane Harvey further demonstrate that safety measures for the construction of chemical plants are inadequate and pose great public risks.[59]

It is a truism of economics that monopolies stifle competition. They resist game-changing technology lest "their lunch be eaten," as the Silicon Valley phrase goes. Think of the great Thomas Edison with his ownership of direct current (DC) electric power trying to stop Nikola Tesla and George Westinghouse from employing their superior alternating current (AC), or General Motors developing and abandoning electric cars in the 1990s.[60]

Those self-serving actions only delayed progress or changed its trajectory, but much more is at stake now. Fossil fuel companies and governments based on them will understandably attempt to stall carbon-neutral approaches, electric cars, and power derived from wind, solar, wave, hydro, and the rest in order to reap the last profits from coal and oil. It is rare for people to willingly leave twenty-dollar bills buried in the ground. But by stalling to enrich the very, very few, most of whom know quite well what they are doing and what the science says, we are as a species fouling our nest, reaping the whirlwind as the Bible says, since the wind patterns are shifting with climate change.[61]

One of our great strengths as humans, however, is self-awareness and resilience. We also have a hardwired love of family and happiness that may yet balance our innate tribalism. (One can say that diversity, intermarriage, and migrations all act to balance our tribalism, too.)[62] Certain forces may get away with a lot, the public be damned, but for how long? The whole of the United States is not Oklahoma, Kansas, or Texas. These states are making the rules for the rest of us right now. And for what, greed? A lot more children will get asthma, a lot more older people will endure environmental cancers, and a lot more people will unnecessarily die if havoc can be wreaked for the next few years. Personally, we really don't understand how people can in effect murder their fellow citizens for money.

The hydrocarbons still in the ground have value, and we wouldn't be surprised if there are certain technical advancements to make them competitive with solar and wind, and perhaps even cleaner to use than they are now. We may even be so lucky as to invent certain industrial-scale methods of drawing down carbon. But we are not there yet and cannot afford to wait. A best-practices policy could be put in place in regard to natural gas drilling and methane releases as well, even if today those protections to our health are shamelessly under attack.

At what point do people push back? When we're facing another BP oil spill, Love Canal, or Category 5 hurricane in Miami, New Orleans, or New York? A levee-busting major flood event in the Mississippi that would be far worse than the great flood of 1927 because of the much larger numbers of people and vulnerable infrastructures in harm's way? Mammoth cyanobacterial poisoning of the Great Lakes?[63] Will such catastrophes convince even recalcitrant folks to re-think—and re-fund—important environmental initiatives quickly? This see-sawing could go on for a while, and it will take millions of people to stop it and some economic counterforce.

It would help enormously if people stopped to look at the record and the predictive power of the science of global warming versus the lies and obfuscations of the climate deniers. Over the past several years the scientists are batting close to a thousand. Based on data, scientists predicted that warmer oceans would fuel stronger hurricanes, and that's exactly what happened when Harvey transformed from a middling tropical depression into a Category 4 hurricane in just fifty-two hours after it encountered a pool of very warm Gulf water. More generally, the frequency of Category 4 and 5 hurricanes is increasing as seas warm. Based on data, scientists predicted that the 2017 hurricane season would be particularly dangerous, and that's exactly what happened. Based on data and elementary chemistry, scientists predicted the earth would get warmer as greenhouse carbon dioxide from the burning of fossil fuels increased in the atmosphere, and that's what happened. Based on data and elementary chemistry, scientists predicted that sea level would rise as temperatures rose, ice melted, and seawater expanded, and that's what happened. The only thing wrong with their predictions is that the scientists have been so cautious (as responsible scientists tend to be) that they have tended to underestimate the rates of climatic change. We could go on about the scientists increasingly getting it right. In contrast, climate change deniers said none of these things would happen and they were dead wrong.

How heavy a hand can a climate-change-denying federal administration exert to hold back the inevitable progress that is already taking place on so many fronts, as we have detailed earlier? America is a relatively free-market capitalist country. Many of our largest corporations have for some time understood what is happening scientifically and have been preparing their organizations for coming changes in sea level rise and climate, all while saving serious amounts of money from taking the soft path of greater energy efficiency, and advancing technologies that have significantly reduced the cost of wind, solar, storage, and renewables. Cities and states have also rapidly banded together in shared strategies including those related to fuel efficiencies, cap and trade, the banning of toxic chemicals, new agricultural approaches, favoring the use of renewables, and investing in advanced corporate and university research along these lines.[64] These city and state forces are very significant, accounting for as much as two-thirds of the entire GDP of the American economy.[65]

They constitute a very big and powerful work-around from federal intransigence and backtracking in the service of a minority industry and a small number of self-serving donors.

Add in the huge companies like Apple, Exelon, Pacific Gas and Electric, Hewlett-Packard, and many others that, for climate reasons, have opted out of retrograde lobbying groups like the U.S. Chamber of Commerce, and it becomes obvious that the pushback against greed has become forward momentum.[66] All this problem-solving and adaptation would be made much easier with a return to a science-based federal administration, but our view is that there is no stopping the momentum already generated and in the works.

Fortunately, strong bipartisan efforts to respond to the economic risks of climate change are gathering steam. Two former Republican secretaries of state—James Baker and George Schultz—and the former Republican secretary of the treasury, Hank Paulson, have joined with California's Democratic governor Jerry Brown, Washington's Democratic governor Jay Inslee, former Democratic secretary of state John Kerry, former Republican mayor of New York City Michael Bloomberg, former energy secretary Ernie Moniz, the CEOs of General Electric, Pacific Gas and Electric, the mayor of Paris, the director of the World Bank, and citizen Leonardo DiCaprio to outline a way forward that addresses climate change in realistic terms. Their focus is both humanitarian, in terms of successful adaptation to increasing global threats, and economic, in recognizing the hugely valuable opportunities that climate mitigation and adaptation represent to America's most forward-thinking companies regardless of politics. There are no (D)s or (R)s after storm names, as John Kerry put it, nor after the immense fires blazing in America's West.[67]

So how much would it cost to prepare America for what is coming? One informed report co-signed by Paulson, Bloomberg, and hedge-fund financier Tom Steyer estimates "that an average of $320 billion in private investment per year is needed through 2050 to build a clean energy economy and achieve the emissions reductions necessary to avoid the worst economic impacts of climate change. These necessary investments would be similar in scale to other major recent investments made by American business, including in computers and software at $350 billion per year over the past decade. Investments in clean energy could yield on average up to $366 billion in savings per year from reductions in spending on fossil fuels. The country would gain over 1 million new jobs by 2030, with utilities, construction, and manufacturing seeing the largest gains. But 270,000 jobs would be lost in coal mining, oil and gas related jobs, primarily in Southern and Mountain states."[68]

This last is why America must pull together as one. People are suffering in those states whose economies have rapidly changed because coal is so significantly underpriced by natural gas, wind, and solar, and they cannot ethically be left behind. Nor can those who work in oil and gas in the decades to come.

This is a moment in American history when great economic opportunity and the alleviation of potentially enormous human suffering due to climate change are within our grasp. Money, some trillions, are at risk; alternatively, enormous amounts could be made and jobs created. The unfolding humanitarian crisis in Puerto Rico is a Katrina-like reminder of an alternative horror to come if we don't act immediately to move toward a carbon-free economy.[69]

Ours is the country that electrified the world, won World War II, put a man on the moon, and invented the Internet, transistor, and personal computer, not to mention little photovoltaic cells that are the backbone of the solar power industry. After World War II our leaders understood how science and technology had, with much blood and strategy, won and protected our freedom. They created the National Science Foundation and a host of additional government organizations, research institutes, and other ways to transfer technical expertise from and to federally funded universities, all based on science, and these things made us rich, the envy of the world.[70]

What happened to us? Our country was captured by a suite of industries that promoted one way of organizing production, one dedicated to nothing but the short-term bottom line. Its champions have tried to sabotage us in other ways. They created a false ideology to protect their interests against threats from the truth, common sense, and the discoveries of science. Their scheme worked great for a while, but people are catching on. The only thing holding us back is the concerted right-wing war on science, facts, knowledge, and common sense. The far Right has distorted free-market capitalism because what they've created with their subsidies and special privileges isn't free market at all. It's outlaw activity in a rigged system. And they cavil at the mere mention of constructive regulation.

It's time to stop the nonsense, see the problems clearly, and work together, politically and culturally, to restore America to the innovative, rational, scientific, and practical business-oriented country we were from Benjamin Franklin and Thomas Jefferson until now. Smart people realize this is true. Those with heart and honest eyes know what's coming. Old people with grandchildren, young people looking into the future, they can all feel it coming—on the wind.

NOTES

Chapter 1. Corn Off the Cob

1 Arthur G. Sulzberger, "As Crop Prices Soar, Iowa Farms Add Acreage," *New York Times*, December 30, 2011, accessed January 15, 2017, http://www.nytimes.com/2011/12/31/us/in-iowa-farmland-expands-as-crop-prices-soar.html; Rob Cook, "Corn Prices by Year," Beef2Live, accessed January 15, 2017, http://beef2live.com/story-corn-prices-year-0-113575.

2 Throughout the chapter, quotations by John Weber are from Weber, personal communications, September 11, 2011, and November 16, 2012.

3 "Corn Prices 'to Set Record in 2011—by a Margin,'" Agrimoney.com, December 24, 2010, accessed January 15, 2017, http://www.agrimoney.com/news/corn-prices-to-set-record-in-2011---by-a-margin--2650.html; for historical fluctuations, see "Prices Received for Corn by Month—United States Chart," *Agricultural Prices*, September 29, 2015, p. 12, accessed January 15, 2017, http://usda.mannlib.cornell.edu/usda/nass/AgriPric/2010s/2015/AgriPric-09-29-2015.pdf. Prices would eventually drop by half from $7.50/bushel to $3.78/bushel in 2015, as noted by Nelson D. Schwartz and Julie Creswell in "A Global Chill in Commodity Demand Hits America's Heartland," *New York Times*, October 23, 2015, accessed January 15, 2017, https://www.nytimes.com/2015/10/25/business/energy-environment/americas-heartland-feels-a-chill-from-collapsing-commodity-prices.html?_r=0.

4 "Cargill," *Wikipedia*, accessed January 15, 2017, https://en.wikipedia.org/wiki/Cargill.

5 "Crop Production 2015 Summary, January 2016," USDA, accessed January 14, 2017, http://www.usda.gov/nass/PUBS/TODAYRPT/cropan16.pdf; "Faqs," Iowa Corn, accessed January 14, 2017, https://www.iowacorn.org/resources/faqs/.

6 "Iowa," *Census of Agriculture 2012*, Census Publications, USDA, accessed January 15, 2017, https://www.agcensus.usda.gov/Publications/2012/Full_Report/Census_by_State/Iowa/; "Faqs," Iowa Corn.

7 "Industry Statistics," Renewable Fuels Association, accessed January 15, 2017, http://ethanolrfa.org/resources/industry/statistics/#1460745352774-cd978516-814c.

8 For more on the debate, see "Food vs. Fuel," *Wikipedia*, accessed January 14, 2017, https://en.wikipedia.org/wiki/Food_vs._fuel.

9 "U.S. Soy Exports Set New Record," United Soybean Board, December 17, 2013, accessed January 14, 2017, http://unitedsoybean.org/article /u-s-soy-exports-set-new-record/. Starting in 2012, China was the number one importer of U.S. agricultural products, but in 2015 Canada edged ahead with Mexico in third place; see "USDA Economic Research Service, Foreign Agricultural Trade, Table 4, for 2015," USDA, accessed January 15, 2017, http://www.ers.usda.gov/data-products/ag-and-food -statistics-charting-the-essentials/agricultural-trade/.

10 "Tile Drainage," *Wikipedia*, accessed July 29, 2017, https://en.wikipedia .org/wiki/Tile_drainage; Will Hoyer, "Agricultural Drainage and Wetlands: Can They Co-exist?," *Iowa Policy Project*, June 2011, accessed January 15, 2017, http://www.iowapolicyproject.org/2011docs/110622 -wetlands.pdf. Hoyer also cites evidence showing that some 90 percent of Iowa marsh and wetlands were drained long ago.

11 John Sawyer, "Anhydrous Ammonia Application to Dry Soils," *Iowa State University Extension and Outreach*, October 28, 2011, accessed January 15, 2017, http://crops.extension.iastate.edu/cropnews/2011/10 /anhydrous-ammonia-application-and-dry-soils.

12 "Chemical Inputs: Fertilizer Use and Markets," USDA, accessed January 15, 2017, http://www.ers.usda.gov/topics/farm-practices-management /chemical-inputs/fertilizer-use-markets.aspx; Lester R. Brown, "Feeding Everyone Well: Raising Cropland Productivity," chap. 7 in Lester Brown, *Eco-Economy: Building an Economy for the Earth* (New York: Norton, 2001), accessed January 15, 2017, http://www.earth-policy.org/mobile /books/eco/eech7_ss3; John M. Shutske, "Using Anhydrous Ammonia Safely on Farms," University of Minnesota Extension, 2005, accessed January 15, 2017, http://www.extension.umn.edu/agriculture/nutrient -management/nitrogen/using-anhydrous-ammonia-safely-on-the -farm/.

13 "Sugary Drinks Portion Cap Rule," *Wikipedia*, accessed January 14, 2017, https://en.wikipedia.org/wiki/Sugary_Drinks_Portion_Cap_Rule; Nicholas Bakalar, "Fructose-Sweetened Beverages Linked to Heart Risks," *New York Times*, April 23, 2009, accessed January 15, 2017, http:// www.nytimes.com/2009/04/23/health/23sugar.html; Nicholas Bakalar, "High Fructose Heart Risks," *New York Times*, April 27, 2015, accessed January 15, 2017, http://well.blogs.nytimes.com/2015/04/27 /high-fructose-heart-risks/.

14 Donnelle Eller, "China to Buy $5.3 B in U.S. Soybeans," *Des Moines Register*, September 24, 2015, accessed January 15, 2017, http://www .desmoinesregister.com/story/money/2015/09/24/china-buy-53b-us -soybeans/72735864/.

15 Lee Rood, "Water Works Shifts How It Disposes of Excess Nitrates," *Des Moines Register*, October 8, 2016, accessed January 15, 2017, http:// www.desmoinesregister.com/story/news/investigations/readers -watchdog/2016/10/08/water-works-shifts-how-disposes-excess -nitrates/91621994/; Donnelle Eller, "Nitrates in the Water May Be More Harmful Than We Thought," *Des Moines Register*, September 30, 2016, accessed January 15, 2017, http://www.desmoinesregister.com /story/money/agriculture/2016/09/29/elevated-nitrates-linked-cancers -birth-defects-environmental-group-says/91228894/; also see Environmental Health Division, Benton Franklin Health District, "What Are Nitrates?" accessed January 15, 2017, http://www.bfhd.wa.gov/info /nitrate-nitrite.php.

16 E. E. Alberts, R. C. Wendt, and R. E. Burwell, "Corn and Soybean Cropping Effects on Soil Losses and C Factors," *Soil Science Society of America Journal* 49 (1985): 721–728; W. C. Lindemann and C. R. Glover, *Nitrogen Fixation by Legumes*, New Mexico State University College of Agriculture and Home Economics, May 1990, accessed January 15, 2017, http://www.csun.edu/~hcbio027/biotechnology/lec10/lindemann.html; F. Salvagiotti et al., "Nitrogen Uptake, Fixation and Response to Fertilizer N in Soybeans: A Review," *Field Crops Research* 108 (2008): 1–13; L. Christianson, M. J. Castellano, and M. J. Helmers, "Nitrogen and Phosphorus Balances in Iowa Cropping Systems: Sustaining Iowa's Soil Resources," Report to the Iowa State Legislature, 2012, accessed May 17, 2017, https://www.legis.iowa.gov/docs/APPS/AR/5CDA06E3 -F230-4E2B-8FE9-CC1F64BE2CA8/IDALS-Final%20Soil%20Nutrient %20Balance%20Report.pdf.

Chapter 2. Food versus Fuel

1 Richard Pirog and Zackary Paskiet, *A Geography of Taste: Iowa's Potential for Developing Place-Based and Traditional Foods* (Ames, IA: Leopold Center for Sustainable Agriculture, 2004); Wayne J. Guglielmo, "Keeping Them on the Farm," *New Jersey Monthly*, October 11, 2010, accessed January 14, 2017, https://njmonthly.com/articles/jersey-living/keeping -them-on-the-farm/.

2 *Geographic Area Series*, vol. 1, part 51 of the *2012 Census of Agriculture: United States Summary and State Data* (Washington, DC: U.S. Government Printing Office, 2014); Michael Pollan, *The Omnivore's Dilemma: A Natural History of Four Meals* (London: Bloomsbury, 2006); John C. Hudson and Christopher R. Laingen, *American Farms, American Food: A Geography of Farms and Food Production in the United States* (Lanham, MD: Lexington Books, 2016).

3 *Geographic Area Series*.

4 Ibid.

5 Robert Wisner, "Soybean Oil and Biodiesel Usage Projections and Balance Sheet," Agricultural Marketing Resource Center, last modified December 21, 2015, accessed January 14, 2017, www.extension.iastate .edu/agdm/crops/outlook/biodeiselbalancesheet.pdf; "About ADM," Archer Daniels Midland Corporation, accessed January 14, 2017, http:// www.adm.com/en-US/worldwide/us/Pages/Facilities.aspx; "#1 Stock to Buy Right Now," *Ycharts*, accessed January 14, 2017, https://ycharts.com /companies/ADM/market_cap; Chris Prentice, Michael Hirtzer, and Karl Plume, "Ethanol Pioneer ADM's Struggle Reflects Deepening Industry Woes," *Reuters Commodities*, February 3, 2016, accessed February 24, 2017, http://www.reuters.com/article/us-archerdaniels-results -ethanol-idUSKCN0VC0CC; Rod Nickel and Chris Prentice, "ADM Cuts Biodiesel Output as Industry Hit by Weak Margins," *Reuters Market News*, February 25, 2015, accessed February 24, 2017, http://www .reuters.com/article/usa-adm-biodiesel-idUSL1N0VZ1W520150225; John Whims, "Pipeline Considerations for Ethanol," AgMRC, Department of Agricultural Economics, Kansas State University, August 2002, accessed February 24, 2017, http://www.agmrc.org/media/cms /ksupipelineethl_8BA5CDF1FD179.pdf.

6 Jacob Bunge and Jesse Newman, "Falling Crude Prices Force Ethanol Makers to Take It on the Chin," *Wall Street Journal*, January 2, 2015, accessed February 24, 2017, https://www.wsj.com/articles/falling-crude -prices-force-ethanol-makers-to-take-it-on-the-chin-1420238100; Prentice, Hirtzer and Plume, "Ethanol Power ADM's Struggle."

7 "Glenville MN," *Poet: Human + Nature*, accessed February 24, 2017, poet .com/glenville; throughout the chapter, all quotations from Kevin Hobbie are from Hobbie, personal communication, September 9, 2011.

8 Carl Zulauf, "US Corn Ethanol Market: Understanding the Past to Assess the Future," Farmdoc Project, University of Illinois at Urbana-Champaign, November 17, 2016, accessed January 14, 2017, http://

farmdocdaily.illinois.edu/2016/11/us-corn-ethanol-market-past-and
-future.html.

9 Sarah C. Davis et al., "Life-Cycle Analysis and the Ecology of Biofuels," *Trends in Plant Science* 14 (2008): 140–146.

10 J. D. Glover et al., "Increased Food and Ecosystem Security via Perennial Grains," *Science* 328 (2010): 1638–1639.

11 David Tilman, Jason Hill, and Clarence Lehman, "Carbon-Negative
Biofuels from Low-Input High-Diversity Grassland Biomass," *Science*
314 (2006): 1598–1600; David Tilman, David Wedin, and Johannes
Knops, "Productivity and Sustainability Influenced by Biodiversity in
Grassland Ecosystems," *Nature* 379 (1996): 718–720.

12 Jason Hill et al., "Environmental, Economic, and Energetic Costs and
Benefits of Biodiesel and Ethanol Biofuels," *Proceedings of the National
Academy of Sciences USA* 103 (2006): 11206–11210.

13 David Tilman et al., "Beneficial Biofuels—The Food, Energy, and En-
vironment Trilemma," *Science* 325 (2009): 270–271; "Food vs. Fuel,"
Wikipedia, accessed January 14, 2017, https://en.wikipedia.org/wiki
/Food_vs._fuel; as discussed in Chapter 1, roughly half of Iowa corn and
soy production goes to animal feeds and other forms of industrial food.
Timothy Searchinger et al., "Use of U.S. Croplands for Biofuels Increases
Greenhouse Gases through Emissions from Land-Use Change," *Science*
319 (2008): 1238–1240.

14 Joseph Fargione et al., "Land Clearing and the Biofuel Carbon Debt,"
Science 319 (2008): 1235–1238.

15 Jerry B. Brown, Rinaldo Brutoco, and James A. Cusumano, "The Hid-
den Costs of Fossil Fuels," chap. 6 in *Freedom from Mideast Oil* (Santa
Barbara, CA: World Business Academy, 2007), accessed February 24,
2017, worldbusiness.org/wp-content/uploads/2012/12/Freedom
_Chapter6.pdf; "Frequently Asked Questions: How Much Carbon Di-
oxide Is Produced When Different Fuels Are Burned?" Independent
Statistics and Analysis, U.S. Energy Information Administration, accessed February 24, 2017, https://www.eia.gov/tools/faqs/faq.cfm?id
=73&t=11.

16 Jason Hill et al., "Climate Change and Health Costs of Air Emissions
from Biofuels and Gasoline," *Proceedings of the National Academy of Sciences USA* 106 (2009): 2077–2082.

17 "Fertilizer," *How Products Are Made*, accessed February 24, 2017, http://
www.madehow.com/Volume-3/Fertilizer.html; "Pesticide," *How Products Are Made*, accessed February 24, 2017, http://www.madehow.com

/Volume-1/Pesticide.html; Justin Gillis, "New Report Urges Western Governments to Reconsider Reliance on Biofuels," *New York Times*, January 28, 2015, accessed February 24, 2017, https://www.nytimes.com /2015/01/29/science/new-report-urges-western-governments-to -reconsider-reliance-on-biofuels.html?_r=0; Robert Bryce, "End the Ethanol Rip-Off," *New York Times*, March 10, 2015, accessed February 24, 2017, https://www.nytimes.com/2015/03/10/opinion/end-the -ethanol-rip-off.html.

Chapter 3. Where's the Food?

1 Iowa leads the nation in ethanol production with 42 plants that produced 3.9 billion gallons in 2014. The renewable fuels industry (including biodiesel) supports 43,000 jobs and accounts for 3.5 percent of Iowa's GDP. Further information can be found at "Ethanol and the Economy," Iowa Renewable Fuels Association, accessed February 24, 2017, iowarfa.org /ethanol-center/ethanol-facts/ethanol-and-the-economy/; and Christopher Doering, "Iowa Produces Record Amount of Ethanol in 2014," *Des Moines Register*, December 29, 2014, accessed February 24, 2017, http:// www.desmoinesregister.com/story/money/agriculture/green-fields /2014/12/29/iowa-ethanol-record-production/21004575/.

2 "Faqs," Iowa Corn, accessed January 14, 2017, https://www.iowacorn .org/resources/faqs/; "Iowa's Rank in United States Agriculture," USDA, accessed February 24, 2017, https://www.nass.usda.gov/Statistics_by _State/Iowa/Publications/Rankings/2015%20Rankings.pdf; "Poultry Production and Value, 2015 Summary, April 2016," USDA, accessed February 24, 2017, http://usda.mannlib.cornell.edu/usda/current /PoulProdVa/PoulProdVa-04-28-2016.pdf; "Iowa Pork Facts," Iowa Pork Producers Association, accessed February 24, 2017, http://www .iowapork.org/news-from-the-iowa-pork-producers-association/iowa -pork-facts/.

3 Rachel Carson, *Silent Spring* (Boston: Houghton Mifflin, 1962).

4 Throughout the chapter, quotations by Marvin Hotz are from Hotz, personal communication, September 12, 2011.

5 "U.S. Corn Production, 1935–2015," *World of Corn*, National Corn Growers Association, January 12, 2016, accessed February 24, 2017, http://www.worldofcorn.com/#us-corn-production. The *World of Corn* site is a shorthand view of the corn world, and is simpler to browse than USDA statistics.

6 Throughout the chapter, quotations by Kandy and Fred Schlichting are from Schlichting and Schlichting, personal communication, September 12, 2011.

7 "Eagle Point Park (Dubuque, Iowa)," *Wikipedia*, last modified September 9, 2016, accessed February 24, 2017, https://en.wikipedia.org/wiki /Eagle_Point_Park_(Dubuque,_Iowa).

8 Calvin R. Fremling, *Immortal River: The Upper Mississippi in Ancient and Modern Times* (Madison: University of Wisconsin Press, 2005), 23.

9 Ibid., 215–229; Robert H. Meade and John A. Moody, "Causes for the Decline of Suspended-Sediment Discharge in the Mississippi River System, 1940–2007," *Hydrological Progress* 24 (2010): 35–49, doi: 10.1002/ hyp.7477.

10 "Towboats and Barges of the Middle Mississippi River Valley," *Visitors Guide to the Middle Mississippi River Valley*, accessed February 24, 2017, http://www.greatriverroad.com/all/towboats.htm; "Pusher (Boat)," *Wikipedia*, last updated November 11, 2016, accessed February 24, 2017, https://en.wikipedia.org/wiki/Pusher_(boat).

Chapter 4. Don't Drink That!

1 Neal Lawrence, "Danger on Tap," *Midwest Today* (Spring 1998), accessed February 24, 2017, http://www.midtod.com/dangerontap.html; E. G. Stets, V. J. Kelly, and C. G. Crawford, "Regional and Temporal Differences in Nitrate Trends Discerned from Long-Term Water Quality Monitoring Data," *Journal of the American Water Resources Association* 51 (2015): 1394–1407, accessed February 24, 2017, doi: 10.1111/1752–1688 .12321; "Glyphosate Herbicide Found in Many Midwestern Streams, Antibiotics Not Common," *Environmental Health—Toxic Substances Hydrology Program*, USGS, accessed February 24, 2017, https://toxics.usgs .gov/highlights/glyphosate02.html.

2 "Cuyahoga River," *Wikipedia*, accessed February 24, 2017, https://en .wikipedia.org/wiki/Cuyahoga_River; Jonathan H. Adler, "Fables of the Cuyahoga: Reconstructing a History of Environmental Protection," *Fordham Environmental Law Journal* 14 (2003): 95–98, 103–104, accessed February 24, 2017, http://scholarlycommons.law.case.edu/cgi /viewcontent.cgi?article=1190&context=faculty_publications; "The Cities: The Price of Optimism," *Time*, August 1, 1969, accessed February 24, 2017, http://content.time.com/time/magazine/article/0,9171,901182,00 .html.

3 "The Guardian: Origins of the EPA," U.S. Environmental Protection Agency, accessed February 25, 2017, https://archive.epa.gov/epa/aboutepa/guardian-origins-epa.html; "History of the Clean Water Act," U.S. Environmental Protection Agency, accessed February 25, 2017, https://www.epa.gov/laws-regulations/history-clean-water-act.

4 James Salzman, "Rivers No Longer Burn," *Slate*, December 10, 2012, accessed February 25, 2017, http://www.slate.com/articles/health_and_science/science/2012/12/clean_water_act_40th_anniversary_the_greatest_success_in_environmental_law.html; Great Lakes Burning River Fest, accessed October 6, 2017, https://burningriverfest.org/; Nadia Arumugam, "Oysters Are Back in New York City Waters," *Forbes*, February 27, 2012, accessed February 25, 2017, http://www.forbes.com/sites/nadiaarumugam/2012/02/27/oysters-are-back-in-new-york-city-waters/#7fac6ec22e07; John Waldman, *Heartbeats in the Muck: A Dramatic Look at the History, Sealife, and Environment of New York Harbor* (New York: Lyons Press, 1999); "Cuyahoga Valley," National Park Service, https://www.nps.gov/cuva/learn/nature/waterquality.htm; "Nonpoint Source Pollution," *Wikipedia*, accessed February 25, 2017, https://en.wikipedia.org/wiki/Nonpoint_source_pollution; R. Eugene Turner and Nancy N. Rabalais, "Linking Landscape and Water Quality in the Mississippi River Basin for 200 Years," *BioScience* 53 (2003): 563–572; Craig Cox and Andrew Hug, "Murky Waters: Farm Pollution Stalls Cleanup of Iowa Streams," *Environmental Working Group* (December 2012): 1–52, accessed February 25, 2017, http://static.ewg.org.s3.amazonaws.com/reports/2012/murky_waters/Murky_Waters.pdf; "Flint Water Crisis," *Wikipedia*, accessed February 25, 2017, https://en.wikipedia.org/wiki/Flint_water_crisis; Josh Sanburn, "Flint's Water Crisis Still Isn't Over: Here's Where Things Stand a Year Later," *Time*, January 18, 2017, accessed February 25, 2017, http://time.com/4634937/flint-water-crisis-criminal-charges-bottled-water/; Donovan Hohn, "Flint's Water Crisis and the 'Troublemaker' Scientist," *New York Times Magazine*, August 16, 2016, accessed February 25, 2017, https://www.nytimes.com/2016/08/21/magazine/flints-water-crisis-and-the-troublemaker-scientist.html.

5 *Census of Agriculture 2012*, Census Publications, USDA, accessed January 15, 2017, https://www.agcensus.usda.gov/Publications/2012/Full_Report/; Linda J. Lear, *Rachel Carson: Witness for Nature* (Boston: Houghton Mifflin Harcourt, 2009).

6 We thank Matt Leibman for pointing out these references: "Nutrients Remain Elevated in the Nation's Groundwater," USGS National Water Quality Assessment Project, modified on March 4, 2014, accessed May 17, 2017, https://water.usgs.gov/nawqa/home_maps/nutrients.html; Neil M. Dubrovsky et al., "Nutrients in the Nation's Streams and Groundwater, 1993–2004," USGS Circular 1350, U.S. Department of the Interior, Reston, VA, 2010, accessed May 17, 2017, https://pubs.usgs.gov /circ/1350/pdf/circ1350.pdf; Fertilizer Institute, accessed May 18, 2017, https://www.tfi.org/; Kathy Mathers, "1980–2014: U.S. Corn Production More than Doubles, While Fertilizer Use Remains Almost Level," Fertilizer Institute, May 14, 2014, accessed February 25, 2017, https:// www.tfi.org/the-feed/1980-2014-us-corn-production-more-doubles -while-fertilizer-use-remains-al; John P. Schmidt, "Nitrogen Fertilizer for Soybean?" *DuPont Pioneer,* accessed February 25, 2017, https://www .pioneer.com/home/site/us/agronomy/library/nitrogen-fertilizer-for -soybean/; Mike Stanton, "Phosphorus and Potassium Fertilizer Rec- ommendations for High-Yielding, Profitable Soybeans," Michigan State University Extension, March 17, 2014, accessed February 25, 2017, http://msue.anr.msu.edu/news/phosphorus_and_potassium_fertilizer _recommendations_for_high_yielding_profi.

7 "Iowa," *Census of Agriculture 2012,* Census Publications, USDA, accessed January 15, 2017, https://www.agcensus.usda.gov/Publications/2012/Full _Report/Census_by_State/Iowa/.

8 *Census of Agriculture 2012.*

9 Jake, May 22, 2016, "How Much Roundup Do Farmers Actually Use?" *A Year in the Life of a Farmer,* accessed February 25, 2017, https:// southsaskfarmer.com/2016/05/22/how-much-roundup-do-farmers -actually-use/; ibid.; "Nutrients Remain Elevated in the Nation's Ground- water."

10 Jerry Adler, "The Growing Menace from Superweeds," *Scientific Ameri- can* (May 2011): 74–79, accessed February 25, 2017, https://www .scientificamerican.com/article/the-growing-menace-from-superweeds /; Marion Nestle, "Superweeds: A Long-Predicted Problem for GM Crops Has Arrived," *Atlantic,* May 15, 2012, accessed February 25, 2017, http://www.theatlantic.com/health/archive/2012/05/superweeds -a-long-predicted-problem-for-gm-crops-has-arrived/257187/; Natasha Gilbert, "A Hard Look at GM Crops," *Nature* 497 (2013): 24–26, accessed February 25, 2017, http://www.nature.com/polopoly_fs/1.12907!/menu

/main/topColumns/topLeftColumn/pdf/497024a.pdf; Bill Spiegel, "Pigweed Problems Plague the Cornbelt," *Successful Farming at Agriculture.com*, July 24, 2016, accessed February 25, 2017, http://www.agriculture.com/crops/corn/pigweed-problems-prevailing.

11 *Superweeds: How Biotech Crops Bolster the Pesticide Industry*, Food and Water Watch, 2013, accessed February 25, 2017, http://www.foodandwaterwatch.org/sites/default/files/Superweeds%20Report%20July%202013.pdf; Mary Turck, "EPA Approves Another Superchemical for Superweeds," *Aljazeera America*, December 12, 2014, accessed February 25, 2017, http://america.aljazeera.com/opinions/2014/12/enlist-duo-superweedsglyphosatemonsantoepa.html; "Glyphosate-Resistant Weed Problem Extends to More Species, More Farms," *Farm Industry News*, January 29, 2013, accessed February 25, 2017, http://www.farmindustrynews.com/ag-technology-solution-center/glyphosate-resistant-weed-problem-extends-more-species-more-farms; William Neuman and Andrew Pollack, "Farmers Cope with Roundup-Resistant Weeds," *New York Times*, May 3, 2010, accessed May 27, 2017, http://www.nytimes.com/2010/05/04/business/energy-environment/04weed.html?_r=0; David A. Mortensen et al., "Navigating a Critical Juncture for Sustainable Weed Management," *BioScience* 62 (2012): 75–84, doi: 10.1525/bio.2012.62.1.12.

12 Mortensen et al., "Navigating a Critical Juncture"; Josh Harkinson, "Farmers Say This Weedkiller Is Also Killing Their Soybean Plants," *Mother Jones*, June 28, 2017, accessed October 1, 2017, http://www.motherjones.com/food/2017/06/farmers-say-this-weedkiller-is-also-killing-their-soybean-plants/; Caitlin Dewey, "This Miracle Weedkiller Was Supposed to Save Farms. Instead, It's Devastating Them," *Washington Post*, August 29, 2017, accessed October 1, 2017, https://www.washingtonpost.com/business/economy/this-miracle-weed-killer-was-supposed-to-save-farms-instead-its-devastating-them/2017/08/29/33a21a56-88e3-11e7-961d-2f373b3977ee_story.html?utm_term=.021250021461; Danny Hakim, "Monsanto's Weed Killer, Dicamba, Divides Farmers," *New York Times*, September 21, 2017, accessed October 1, 2017, https://www.nytimes.com/2017/09/21/business/monsanto-dicamba-weed-killer.html?mcubz=0&_r=0.

13 Carson, *Silent Spring*.

14 Emma G. Fitzsimmons, "Tap Water Ban for Toledo Residents," *New York Times*, August 3, 2014, accessed February 25, 2017, https://www.nytimes.com/2014/08/04/us/toledo-faces-second-day-of-water-ban.html?_r=0; John Seewer, "Toledo Better Prepared to Keep Toxins Out

of Tap Water," *Detroit News*, July 26, 2015, accessed February 25, 2017, http://www.detroitnews.com/story/news/local/michigan/2015/07/26 /toledo-better-prepared-keep-toxins-tap-water/30696495/; "Water Still Not Safe to Drink in Toledo," *Columbus Dispatch*, August 4, 2014, accessed February 25, 2017, http://www.dispatch.com/content/stories /local/2014/08/04/lake-algae-lurks-as-city-awaits-word.html; Michael Wines, "Behind Toledo's Water Crisis, a Long-Troubled Lake Erie," *New York Times*, August 4, 2014, https://www.nytimes.com/2014/08/05 /us/lifting-ban-toledo-says-its-water-is-safe-to-drink-again.html.

15 "Microcystin-LR," *Wikipedia*, January 19, 2017, accessed February 25, 2017, https://en.wikipedia.org/wiki/Microcystin-LR; Lesley V. D'Anglada and Jamie Strong, "Drinking Water Health Advisory for the Cyanobacterial Microcystin Toxins," document no. 820R15100, U.S. Environmental Protection Agency, June 15, 2015, accessed February 24, 2017, https://www.epa.gov/sites/production/files/2015-06/documents /microcystins-report-2015.pdf.

16 Jim Erickson, "Record-Breaking 2011 Lake Erie Algae Bloom May Be Sign of Things to Come," *Michigan News*, University of Michigan, April 1, 2013, accessed February 24, 2017, http://www.ns.umich.edu/new /releases/21342-record-breaking-2011-lake-erie-algae-bloom-may-be -sign-of-things-to-come; Anna M. Michalak et al., "Record-Setting Algal Bloom in Lake Erie Caused by Agricultural and Meteorological Trends Consistent with Expected Future Conditions," *Proceedings of the National Academy of Sciences USA* 110 (2013): 6448–6452, accessed February 25, 2017, doi: 10.1073/pnas.1216006110. For an excellent review of the nutrient pollution of Lake Erie and the Great Lakes in general, see Dan Egan, "North America's 'Dead' Sea," chap. 7 of Dan Egan, *The Death and Life of the Great Lakes* (New York: Norton, 2017), 212–244.

17 Ingrid Chorus and Jamie Bartram, *Toxic Cyanobacteria in Water: A Guide to Their Public Health Consequences, Monitoring, and Management* (London: E & FN Spon, 1999), accessed February 24, 2017, http://www .plannacer.msal.gov.ar/images/stories/ministerio/intoxicaciones /cianobacterias/toxcyanobacteria.pdf.

18 James J. Elser, "The Pathway to Noxious Cyanobacteria Blooms in Lakes: The Food Web as the Final Turn," *Freshwater Biology* 42 (1999): 537–543, accessed February 25, 2017, doi: 10.1046/j.1365-2427.1999.00471.x. The invasion of the zebra mussel into the Great Lakes is treated in detail in Dan Egan, "Noxious Cargo," chap. 4 of Egan, *Death and Life of the Great Lakes*, 108–147.

19 Michalak et al., "Record-Setting Algal Bloom in Lake Erie."

20 Craig Cox, Andrew Hug, and Nils Bruzelius, "Losing Ground," Environmental Working Group, April 2011, accessed February 25, 2017, http://static.ewg.org/reports/2010/losingground/pdf/losingground _report.pdf; John Eligon, "After Drought, Rains Plaguing Midwest Farms," *New York Times*, June 9, 2013, accessed February 25, 2017, http://www.nytimes.com/2013/06/10/us/after-drought-rains-plaguing -midwest-farms.html; Allan Konopka and Thomas D. Brock, "Effect of Temperature on Blue-Green Algae (Cyanobacteria) in Lake Mendota," *Applied and Environmental Microbiology* 36 (1978): 572–576, accessed February 25, 2017, http://aem.asm.org/content/36/4/572.full.pdf; Tamar Zohary and Charles M. Breen, "Environmental Factors Favouring the Formation of *Microcystis aeruginosa* Hyperscums in a Hypertrophic Lake," *Hydrobiologia* 178 (1989): 179–192, accessed February 25, 2017, doi: 10.1007/BF00006025; Cayelan C. Carey et al., "Eco-Physiological Adaptations That Favour Freshwater Cyanobacteria in a Changing Climate," *Water Research* 46 (2012): 1394–1407, accessed February 26, 2017, doi: 10.1016/j.watres.2011.12.016; Michalak et al., "Record-Setting Algal Bloom in Lake Erie."

21 Henry A. Vanderploeg et al., "Zebra Mussel (*Dreissena polymorpha*) Selective Filtration Promoted Toxic *Microcystis* Blooms in Saginaw Bay (Lake Huron) and Lake Erie," *Canadian Journal of Fisheries and Aquatic Science* 58 (2001): 1208–1221, accessed February 25, 2017, doi: 10.1139/ cjfas-58-6-1208; David F. Raikow et al., "Dominance of the Noxious Cyanobacterium *Microcystis aeruginosa* in Low-Nutrient Lakes Is Associated with Exotic Zebra Mussels," *Limnology and Oceanography* 49 (2004): 482–487.

22 Olga V. Naidenko, Craig Cox, and Nils Bruzelius, "Troubled Waters: Farm Pollution Threatens Drinking Water," Environmental Working Group, April 12, 2012, accessed February 25, 2017, http:/www.ewg.org /report/troubled waters; Ian Falconer et al., "Safe Levels and Safe Practices," chap. 5 in Ingrid Chorus and Jamie Bartram, eds., *Toxic Cyanobacteria in Water: A Guide to Their Public Health Consequences, Monitoring, and Management* (London: E & FN Spon, 1999), accessed February 25, 2017, http://www.who.int/water_sanitation_health/resourcesquality /toxcyanchap5.pdf.

23 Data downloaded on October 5, 2011, by the Environmental Working Group from "USGS Water Information for the Nation," USGS National Water Information System Web Interface, last modified on Feb-

ruary 25, 2017, accessed February 25, 2017, https://waterdata.usgs.gov
/nwis; "What Is the Regulatory Status of HABs, Cyanobacteria and
Cyanotoxins in the U.S.?," Nutrient Policy and Data, Guidelines and
Recommendations, accessed February 25, 2017, https://www.epa.gov
/nutrient-policy-data/guidelines-and-recommendations#what1; "Cyanobacteria and Cyanotoxins: Information for Drinking Water Systems,"
U.S. Environmental Protection Agency, EPA-810F11001, September 2014, accessed February 25, 2017, https://www.epa.gov/sites
/production/files/2014-08/documents/cyanobacteria_factsheet.pdf; Jennifer L. Graham et al., "Environmental Factors Influencing Microcystin Distribution and Concentration in the Midwestern United States,"
Water Research 38 (2004): 4395–4404, accessed February 25, 2017, doi:
10.1016/j.watres.2004.08.004; Naidenko, Cox, and Bruzelius, "Troubled
Waters."

24 Naidenko, Cox, and Bruzelius, "Troubled Waters"; data downloaded on
October 5, 2011, by the Environmental Working Group from "USGS
Water Information for the Nation"; "Table of Regulated Drinking
Water Contaminants," Ground Water and Drinking Water, U.S. Envi-
ronmental Protection Agency, accessed February 25, 2017, https://www
.epa.gov/ground-water-and-drinking-water/table-regulated-drinking
-water-contaminants#Inorganic; Margaret McCasland et al., "Nitrate:
Health Effects in Drinking Water," Pesticide Safety Education Program
(PSEP), Natural Resources, Cornell University Cooperative Extension,
accessed February 25, 2017, http://psep.cce.cornell.edu/facts-slides-self
/facts/nit-heef-grw85.aspx; Mary H. Ward et al., "Nitrate Intake and the
Risk of Thyroid Cancer and Thyroid Disease," *Epidemiology* 21 (2010):
389–395, doi: 10.1097/EDE.0b013e3181d6201d.

25 Kathryn Z. Guyton et al., "Carcinogenicity of Tetrachlorvinphos, Para-
thion, Malathion, Diazinon, and Glyphosate," *Lancet: Oncology* 16
(May 2015): 490–491, accessed February 25, 2017, doi: http://dx.doi.org
/10.1016/S1470-2045(15)70134-8; "Joint FAO/WHO Meeting on Pesticide Residues: Summary Report," Food and Agriculture Organization
of the United Nations, May 16, 2016, accessed February 25, 2017, http://
www.who.int/foodsafety/jmprsummary2016.pdf?ua=1; Arthur Neslen,
"Glyphosate Unlikely to Pose Risk to Humans, UN/WHO Study Says,"
Guardian, May 16, 2016, accessed February 25, 2017, https://www
.theguardian.com/environment/2016/may/16/glyphosate-unlikely-to
-pose-risk-to-humans-unwho-study-says; Meriel Watts et al., "Glyphosate," PAN: Pesticide Action Network International, October 2016,

accessed February 25, 2017, http://library.ipamglobal.org/jspui/bitstream /ipamlibrary/885/1/Glyphosate-monograph.pdf; Kiera Butler, "A Scientist Didn't Disclose Important Data—and Let Everyone Believe a Popular Weedkiller Causes Cancer," *Mother Jones*, June 15, 2017, accessed November 25, 2017, http://www.motherjones.com/environment /2017/06/monsanto-roundup-glyphosate-cancer-who/; Arthur Neslen, "EU Report on Weedkiller Safety Copied Text from Monsanto Study," *Guardian*, September 14, 2017, downloaded November 25, 2017, https:// www.google.com/search?q=eu+report+on+weedkiller+safety+copied&oq =eu+report+on+weedkiller+safety+copied&aqs=chrome..69i57j0j69i6012 .11650j0j7&sourceid=chrome&ie=UTF-8.

26 A. Roustan et al., "Genotoxicity of Mixtures of Glyphosate and Atrazine and Their Environmental Transformation Products before and after Photoactivation," *Chemosphere* 108 (August 2014): 93–100, accessed February 25, 2017, doi.org/10.1016/j.chemosphere.2014.02.079.

27 Arthur Neslen, "Controversial Chemical in Roundup Weedkiller Escapes Immediate Ban," *Guardian*, June 29, 2016, accessed February 25, 2017, https://www.theguardian.com/business/2016/jun/29/controversial -chemical-roundup-weedkiller-escapes-immediate-ban; Tania Rabesandratana, "Europe Stalls Weed Killer Renewal, Again," *Science: ScienceInsider*, May 20, 2016, accessed February 15, 2017, doi: 10.1126/science .aago553; ibid.; PHYS ORG, "EU Parliament Votes to Ban Controversial Weedkiller by 2022, October 24, 2017, downloaded November 25, 2017, https://phys.org/print428059783.html; Simon Marks, Florian Eder, and Giulia Paravicini, "Diplomats: EU Reapproves Glyphosate for Five Years," Politico, November 27, 2017, downloaded January 8, 2018, https://www .politico.eu/article/diplomats-eu-reapproves-glyphosate-for-five-years/.

28 Jennifer Sass and Nina Hwang, "Overview: Regulatory Review of Glyphosate," National Resources Defense Council, November 12, 2015, accessed February 25, 2017, https://www.nrdc.org/sites/default/files/hea _15111002a.pdf; "Glyphosate Issue Paper: Evaluation of Carcinogenic Potential," Office of Pesticide Programs, U.S. Environmental Protection Agency, September 12, 2016, accessed February 25, 2017, https:// www.epa.gov/sites/production/files/2016-09/documents/glyphosate _issue_paper_evaluation_of_carcincogenic_potential.pdf; Associated Press, "California Gets Closer to Requiring Cancer Warning Label on Roundup Weed Killer," *Los Angeles Times*, February 11, 2015, accessed February 25, 2017, http://www.latimes.com/business/la-fi-roundup -cancer-20170127-story.html; "Monsanto, California Battle over Listing

Glyphosate as a Carcinogen," *Food Democracy Now*, January 19, 2017, accessed February 25, 2017, http://www.fooddemocracynow.org/blog/2017/jan/19-0; Karl Plume, "Monsanto Sues to Keep Herbicide off California List of Carcinogens," *Reuters: Business News*, January 21, 2016, accessed February 25, 2016, http://www.reuters.com/article/us-usa-monsanto-glyphosate-idUSKCN0UZ2RN; Lorraine Chow, "Judge Blocks Monsanto's Bid to Stop California from Listing Glyphosate as Carcinogenic," *EcoWatch*, January 30, 2017, accessed February 25, 2017, http://www.ecowatch.com/california-glyphosate-kennedy-2225578913.html; Margot Roosevelt, "California's Right to Exceed Federal Auto Emissions Standards Is Upheld," *Los Angeles Times*, May 2, 2011, accessed February 25, 2017, http://articles.latimes.com/2011/may/02/local/la-me-clean-cars-20110502; Associated Press, "California Passes Sweeping Auto Emissions Standards," *Fox News: Politics*, January 28, 2012, accessed February 25, 2017, http://www.foxnews.com/politics/2012/01/28/california-passes-sweeping-auto-emission-standards.html.

29 Charles M. Benbrook, "Trends in Glyphosate Herbicide Use in the United States and Globally," *Environmental Sciences Europe* 28, no. 3 (2016), doi: 10.1186/s12302-016-0070-0.

30 Carson, *Silent Spring;* Simon G. Potts et al., "Global Pollinator Declines: Trends, Impacts and Drivers," *Trends in Ecology and Evolution* 25, no. 6 (2010): 345–353, accessed February 25, 2017, doi: 10.1016/j.tree.2010.01.007; Adam J. Vanbergen and the Insect Pollinators Initiative, "Threats to an Ecosystem Service: Pressures on Pollinators," *Frontiers in Ecology and the Environment* 11, no. 5 (2013): 251–259, accessed February 25, 2017, doi: 10.1890/120126; John M. Pleasants and Karen S. Oberhauser, "Milkweed Loss in Agricultural Fields Because of Herbicide Use: Effect on the Monarch Butterfly Population," *Insect Conservation and Diversity* 6, no. 2 (2013): 35–144, accessed February 25, 2017, doi: 10.1111/j.1752–4598.2012.00196.x; "Monarch Butterfly," *Wikipedia*, accessed July 29, 2017, https://en.wikipedia.org/wiki/Monarch_butterfly; Anurag Agrawal, *Monarchs and Milkweed: A Migrating Butterfly, a Poisonous Plant, and Their Remarkable Story of Coevolution* (Princeton, NJ: Princeton University Press, 2017); Penelope R. Whitehorn et al., "Neonicotinoid Pesticide Reduces Bumble Bee Colony Growth and Queen Production," *Science* 336 (2012): 351–352, accessed February 25, 2017, doi: 10.1126/science.1215025; Maj Rundlöf et al., "Seed Coating with a Neonicotinoid Insecticide Negatively Affects Wild Bees," *Nature* 521 (2015): 77–80, accessed February 25, 2017, doi: 10.1038/nature14420; Tom Philpott, "The EPA Finally Admitted

That the World's Most Popular Pesticide Kills Bees—20 Years Too Late," *Mother Jones,* January 7, 2016, accessed February 25, 2017, http://www.motherjones.com/tom-philpott/2016/01/epa-finds-major-pesticide-toxic-bees; Erik Stokstad, "European Bee Study Fuels Debate over Pesticide Ban," *Science* 356 (June 30, 2017), doi: 10.1126/science .356.6345.1321; Jeremy T. Kerr, "A Cocktail of Poisons," *Science* 356 (June 30, 2017), 1331–1332, doi: 10.1126/scienceaan6173; N. Tsvetkov et al., "Chronic Exposure to Neonicotinoids Reduces Honey Bee Health near Corn Crops," *Science* 356 (June 30, 2017): 1395–1397, doi: 10.1126/science.qqm7470; B. A. Woodcock et al., "Country-Specific Effects of Neonicotinoid Pesticides on Honey Bees and Wild Bees," *Science* 356 (June 30, 2017): 1393–1395, doi: 10.1126/science.aaa1190.

31 Garrett Hardin, "The Tragedy of the Commons," *Science* 162 (1968): 1243–1248, accessed February 25, 2017, doi: 10.1126/science.162.3859.1243; Naomi Oreskes and Erik M. Conway, *Merchants of Doubt* (New York: Bloomsbury, 2010).

32 Grant Rodgers, "Supreme Court Case Is Key to Recouping Damages from Nitrate Pollution," *Des Moines Register,* September 13, 2016, accessed February 25, 2017, http://www.desmoinesregister.com/story/news /crime-and-courts/2016/09/13/supreme-court-case-key-recouping -damages-nitrate-pollution/90260770/; Grant Rodgers and Donnelle Eller, "Environmentalists Say Ruling Could Slow Water Quality Efforts," *Des Moines Register,* January 27, 2017, accessed February 25, 2017, http://www.desmoinesregister.com/story/news/crime-and-courts/2017 /01/27/court-des-moines-water-works-cannot-win-damages-nitrate-case /97080760/; Naidenko, Cox, and Bruzelius, "Troubled Waters."

Chapter 5. Delta Dawn

1 Throughout the chapter, quotations by R. Eugene "Gene" Turner are from Turner, personal communications, January 8–10, 2012.

2 Richard Campanella, *Bienville's Dilemma: A Historical Geography of New Orleans* (Lafayette: University of Louisiana, 2008); Rebecca Solnit and Rebecca Snedeker, *Unfathomable City: A New Orleans Atlas* (Berkeley: University of California Press, 2013); Calvin R. Fremling, *Immortal River: The Upper Mississippi in Ancient and Modern Times* (Madison: University of Wisconsin Press, 2005).

3 Fremling, *Immortal River;* John M. Barry, *Rising Tide: The Great Mississippi Flood of 1927 and How It Changed America* (New York: Simon and Schuster, 1998).

4 Michael D. Blum and Harry H. Roberts, "The Mississippi Delta Region: Past, Present, and Future," *Annual Reviews of Earth and Planetary Sciences* 40 (2012): 655–683, doi: 10.1146/annurev-earth-042711-105248.

5 Fremling, *Immortal River.*

6 Blum and Roberts, "Mississippi Delta Region."

7 Ibid.

8 Ibid.; Harry H. Roberts, "Dynamic Changes of the Holocene Mississippi River Delta Plain: The Delta Cycle," *Journal of Coastal Research* 13 (1997): 605–627.

9 For a detailed and entertaining history of the capture of the Red River by the Mississippi, Shreve's Cut, the Old River oxbow, and the construction of the Old River Control Structure by the Army Corps of Engineers, see John McPhee, "The Control of Nature: Atchafalaya," *New Yorker*, February 23, 1987, accessed May 27, 2017, http://www.newyorker.com/magazine/1987/02/23/atchafalaya; also reprinted in his book *The Control of Nature* (New York: Farrar, Straus, and Giroux, 1989), 3–92.

10 Ibid.

11 Steve Chapple, "River Keeper," *Reader's Digest* (May 2010), accessed May 28, 2017, https://issuu.com/belowthesurface/docs/riverkeeper. Other popular environmental articles for *Reader's Digest* include "Into the Eye of a Hurricane" (September 2009), "Reefs at Risk" (August 2009), and "New West" (October 2009).

12 McPhee, "Control of Nature"; "Oroville Dam Crisis," *Wikipedia*, accessed May 26, 2017, https://en.wikipedia.org/wiki/Oroville_Dam_crisis; John Wind and Ray Dums, "Water Usage Considerations," National Petroleum Council Future Transportation Fuels (FTF) Topic Paper no. 31, August 1, 2012, accessed May 27, 2017, http://www.npc.org/FTF_Topic_papers/31-Water_Usage.pdf.

13 "Industry Sectors," Louisiana Mid-Continent Oil and Gas Association, accessed May 19, 2017, http://www.lmoga.com/industry-sectors/.

14 Wesley James, Chunrong Jia, and Satish Kedia, "Uneven Magnitude of Disparities in Cancer Risks from Air Toxics," *International Journal of Environmental Research and Public Health* 9 (2012): 4365–4385, doi: 10.3390/ijerph912465. For a vivid visual picture of "Cancer Alley," see Richard Misrach and Kate Orff, *Petrochemical America* (New York: Aperture, 2014), or view images from the book on their website, accessed May 28, 2017, http://aperture.org/shop/petrochemical-america-richard-misrach-kate-orff-book-3123; Gwen Ottinger, Ellen Griffith Spears, and Kate Orff, "Petrochemical America, Petrochemical Addiction," *Southern*

Spaces, November 26, 2013, accessed May 28, 2017, https://southernspaces .org/2013/petrochemical-america-petrochemical-addiction; Trymaine Lee and Matt Black, "'Cancer Alley' Big Industry, Big Problems: Clusters of Poverty and Sickness Shadow America's Industrial South," *Geography of Poverty*, MSNBC 2015, accessed May 11, 2017, http://www .msnbc.com/interactives/geography-of-poverty/se.html; "Cancer Alley," *Wikipedia*, accessed May 11, 2017, https://en.wikipedia.org/wiki/Cancer _Alley; "St. Gabriel, Louisiana," *Wikipedia*, accessed May 11, 2017, https://en.wikipedia.org/wiki/St._Gabriel,_Louisiana; J. Michael Kennedy, "'Chemical Corridor': By 'Old Man River,' New Health Fear," *Los Angeles Times*, May 9, 1989, accessed May 11, 2017, http://articles.latimes .com/1989-05-09/news/mn-2999_1_kay-gaudet-louisiana-s-mississippi -river-cancer-alley; James, Jia, and Kedia, "Uneven Magnitude."

15 S. P. Tsai et al., "Mortality Patterns among Residents in Louisiana's Industrial Corridor, USA, 1970–99," *Occupational Environmental Medicine* 61 (2004): 295–304, doi: 10.1136/oem.2003.007831.

16 James, Jia, and Kedia, "Uneven Magnitude"; Kennedy, "Chemical Corridor"; Julie Dermansky, "One Community's Fight for Clean Air in Louisiana's Cancer Alley," *Desmog*, April 19, 2017, accessed May 11, 2017, https://www.desmogblog.com/2017/04/19/st-john-baptist-parish-fight -clean-air-louisiana-cancer-alley-denka-chloroprene.

17 "*Taxodium distichum*," *Wikipedia*, last updated May 10, 2017, accessed May 28, 2017, https://en.wikipedia.org/wiki/Taxodium_distichum; William Souder, "How Two Women Ended the Deadly Feather Trade," *Smithsonian Magazine*, March 2013, accessed May 28, 2017, http://www .smithsonianmag.com/science-nature/how-two-women-ended-the -deadly-feather-trade-23187277/.

18 Michael S. Kearney, J. C. Alexis Riter, and R. Eugene Turner, "Freshwater River Diversions for Marsh Restoration in Louisiana: Twenty-Six Years of Changing Vegetative Cover and Marsh Area," *Geophysical Research Letters* 38 (2011): L16405, doi: 10.1029/2011GL047847. For background on the project, see Louisiana Department of Natural Resources, *Caernarvon Freshwater Diversion Project*, September 2003, accessed October 1, 2017, https://www.lacoast.gov/reports/project/3890870~1.pdf.

19 Charles H. Peterson et al., "Long-Term Ecosystem Response to the *Exxon Valdez* Oil Spill," *Science* 302 (2003): 2082–2086, accessed May 28, 2017, www.afsc.noaa.gov/publications/misc_pdf/peterson.pdf.

20 J. B. C. Jackson et al., "Ecological Effects of a Major Oil Spill on Panamanian Coastal Marine Communities," *Science* 243 (1989): 37–44, ac-

cessed May 28, 2017, http://biogeodb.stri.si.edu/oilspill/en/resource
/data/docs/jackson_1989_Ecological_effects_of_a_major_oil_spill.pdf.

21 John Carey, "Louisiana Wetlands Tattered by Industrial Canals, Not
Just River Levees," *Scientific American*, December 1, 2013, accessed
May 28, 2017, https://www.scientificamerican.com/article/carey-louisiana
-wetlands-tattered-by-industrial-canals/; see also "Man Made Threats"
in "Mississippi River Delta," *Wikipedia*, last updated May 14, 2017, accessed May 28, 2017, https://en.wikipedia.org/wiki/Mississippi_River
_Delta; John Tibbets, "Louisiana Wetlands: A Lesson in Nature Appre-
ciation," *Environmental Health Perspectives* 114 (2006): A40–A43,
PMC1332684, accessed May 28, 2017, https://www.ncbi.nlm.nih.gov
/pmc/articles/PMC1332684/; and U.S. Department of the Interior, *The
Impact of Federal Programs on Wetlands*, vol. 2: *A Report to Congress. By the
Secretary of the Interior* (Washington, DC: Government Printing Office,
1994). While a link to this book is no longer available on the USDI website, it can be found at Google Books, accessed May 28, 2017, https://books
.google.com/books?id=SjvW4vf_y_EC&printsec=frontcover&source
=gbs_ge_summary_r&cad=0#v=onepage&q&f=false.

22 Nancy Rabalais, personal communication, July 7, 2011.

23 R. Eugene Turner and Nancy N. Rabalais, "Coastal Eutrophication near
the Mississippi River Delta," *Nature* 368 (1994): 619–621, accessed May 28,
2017, https://www.researchgate.net/profile/N_Rabalais/publication
/31951125_Coastal_Eutrophication_Near_the_Mississippi_River
_Delta/links/53ecebeaocf2981ada110320/Coastal-Eutrophication-Near
-the-Mississippi-River-Delta.pdf.

24 James Owen, "World's Largest Deadzone Suffocating Sea," *National Geographic News*, March 6, 2010, accessed May 28, 2017, http://news.national
geographic.com/news/2010/02/100305-baltic-sea-algae-dead-zones
-water/; National Oceanic and Atmospheric Administration (NOAA),
"What Is a Deadzone?," accessed May 28, 2017, http://oceanservice.noaa
.gov/facts/deadzone.html; Chelsea Harvey, "Scientists Predict a Gulf of
Mexico 'Dead Zone' the Size of New Jersey This Summer," *Washington
Post*, June 21, 2017, accessed October 1, 2017, https://www.washingtonpost
.com/news/energy-environment/wp/2017/06/21/scientists-predict-a-gulf
-of-mexico-dead-zone-the-size-of-new-jersey-this-summer/?utm_term=
.9ea07d104163; Jenna Gallegos, "The Gulf of Mexico Dead Zone Is
Larger Than Ever. Here's What to Do about It," *Washington Post*, August 4, 2017, accessed October 1, 2017, https://www.washingtonpost.com
/news/energy-environment/wp/2017/08/04/gulf-of-mexico-dead-zone

-is-larger-than-ever-heres-what-to-do-about-it/?utm_term=.4979ff90586e;
NOAA, "Gulf of Mexico Dead Zone Is Largest Ever Measured," August 4, 2017, accessed October 1, 2017, http://www.noaa.gov/media
-release/gulf-of-mexico-dead-zone-is-largest-ever-measured.

25 Blum and Roberts, "Mississippi Delta Region."

26 "Diatom," *Wikipedia*, last updated May 6, 2017, accessed May 28, 2017, https://en.wikipedia.org/wiki/Diatom; Jack Hall, "The Most Important Organism," *Ecology Global Network*, September 12, 2011, accessed May 28, 2017, http://www.ecology.com/2011/09/12/important-organism/.

27 Hall, "Most Important Organism"; Turner and Rabalais, "Coastal Eutrophication near the Mississippi River Delta."

28 Nancy N. Rabalais et al. "Sediments Tell the History of Eutrophication and Hypoxia in the Northern Gulf of Mexico," *Ecological Applications* 17, special issue (2007): S129–S143, doi: 10.1890/06-0644.1.

29 B. K. Sen Gupta, R. E. Turner, and N. N. Rabalais, "Seasonal Oxygen Depletion in Continental-Shelf Waters of Louisiana: Historical Record of Benthic Foraminifers," *Geology* 24 (1996): 227–230, doi: 10.1130/0091
-7613(1996)024<0227:SODICS>2.3.CO;2.

Chapter 6. Blue Bayou

1 Woodland Plantation & Spirits Hall, accessed May 28, 2017, http://www
.woodlandplantation.com/.

2 Throughout the chapter, all quotations by Foster Creppel are from Creppel, personal communication, November 5, 2011.

3 Ibid.; also see Woodland Plantation & Spirits Hall; "Southern Comfort," *Wikipedia*, updated May 11, 2017, accessed May 28, 2017, https://en
.wikipedia.org/wiki/Southern_Comfort.

4 Woodland Plantation & Spirits Hall.

5 David Simon, *Treme*, HBO series broadcast 2010–2013, accessed May 28, 2017, http://www.hbo.com/treme.

6 The Columns Hotel, accessed May 28, 2017, http://www.thecolumns
.com/; National Park Service, *Jean Lafitte National Park and Preserve, Louisiana*, accessed May 28, 2017, https://www.nps.gov/jela/index.htm.

7 Coastal Protection and Restoration Authority, "Lake Hermitage Marsh Creation," *Types of Projects*, accessed May 21, 2017, http://coastal.la.gov
/project/lake-hermitage-marsh-creation; Mark Schleifstein, "Coal, Petroleum Coke Debris Found in Plaquemines Marsh Restoration Projects," *Times-Picayune*, September 2, 2014, accessed May 21, 2017,

http://www.nola.com/environment/index.ssf/2014/09/coal_petroleum
_coke_debris_fou.html.

8 Amanda Adolph, personal communication, July 9, 2011.

9 Allison Flyer, "Facts for Features: Katrina Impact," Data Center, August 26, 2016, accessed May 28, 2017, http://www.datacenterresearch.org
/data-resources/katrina/facts-for-impact/; "Study: 1927 Mississippi
Flood Would Cost up to $160 Billion," *Insurance Journal*, May 18, 2007,
accessed May 28, 2017, http://www.insurancejournal.com/news/national
/2007/05/18/79826.htm.

10 John M. Barry, *Rising Tide: The Great Mississippi Flood of 1927 and How It
Changed America* (New York: Simon & Schuster, 1997), 238–258, 346–
360; for more on the Great Mississippi Flood, see also Susan Scott Parrish, *The Flood Year 1927: A Cultural History* (Princeton, NJ: Princeton
University Press, 2017).

11 John Lopez et al., *Bohemia Spillway in Southeastern Louisiana: History,
General Description, and 2011 Hydrologic Surveys* (New Orleans: Lake
Pontchartrain Basin Foundation, 2013): 9, accessed May 28, 2017, http://
saveourlake.org/wp-content/uploads/PDF-Documents/our-coast
/Bohemia/Bohemia%20Report_March2013.pdf.

12 Ezra Boyd, "Fatalities Due to Hurricane Katrina's Impacts in Louisiana," Ph.D. diss., University of New Orleans, 2011.

13 Michael D. Blum and Harry H. Roberts, "The Mississippi Delta Region: Past, Present, and Future," *Annual Reviews of Earth and Planetary
Sciences* 40 (2012): 655–683, doi: 10.1146/annurev-earth-042711-105248;
Robert H. Meade and John A. Moody, "Causes for the Decline of
Suspended-Sediment Discharge in the Mississippi River System, 1940–
2007," *Hydrological Progress* 24 (2010): 35–49, doi: 10.1002/hyp.7477.

14 Throughout the chapter, quotations by John Lopez are from Lopez,
personal communication, July 10, 2011.

15 For an interesting discussion of Mississippi sediment load caused by historical change in Midwestern land-use practice, see Andrew W. Tweel and
R. Eugene Turner, "Watershed Land Use and River Engineering Drive
Wetland Formation and Loss in the Mississippi River Birdfoot Delta,"
Limnology and Oceanography 57 (2012): 18–28, doi: 10.4319/lo.2012.57.1.0018.

16 "2012 BCS National Championship Game," *Wikipedia*, accessed May 28,
2017, https://en.wikipedia.org/wiki/2012_BCS_National_Championship
_Game.

17 Throughout the chapter, quotations by John Tesvich are from Tesvich,
personal communication, January 9, 2012.

18 *Fastline Sales,* accessed May 22, 2017, https://www.fastline.com/v100 /listings.aspx?Category=Tractors&Manufacturer=John+Deere&Model =8335R%2c8335RT&HorsePower=4.

19 "Louisiana Oyster Task Force," Department of Wildlife and Fisheries, accessed May 28, 2017, http://www.wlf.louisiana.gov/fishing/oyster -task-force-fishery-information.

20 Throughout the chapter, quotations by Erik Hansen are from Hansen, personal communication, January 10, 2012.

21 Louisiana Department of Wildlife and Fisheries, *Louisiana Shrimp Commercial Rules and Regulations,* accessed May 28, 2017, http://www.wlf .louisiana.gov/sites/default/files/pdf/page/37779-fisheries-brochures /shrimpbrochure20169-24-15low.pdf; "Louisiana Shrimp Fishery," Louisiana Fisheries Forward, updated May 18, 2017, accessed May 28, 2017, http://www.lafisheriesforward.org/fisheries/shrimp/.

22 Mark Schleifstein, "As One Legislator Tears Up, Senate Committee OKs Louisiana Coastal Master Plan," *Times-Picayune,* May 11, 2017, accessed May 12, 2017, http://www.nola.com/environment/index.ssf/2017 /05/coast_master_plan_budget_appro.html.

23 Office of the Governor, Coastal Protection and Restoration Authority, accessed May 12, 2017, http://coastal.la.gov/; Coastal Protection and Restoration Authority, *Types of Projects.* This is on top of the existing 3,500-mile levee system on the Mississippi and its tributaries (MR&T) operated and maintained by the Army Corp of Engineers, which is described in "Levee Systems," U.S. Army Corps of Engineers, Mississippi Valley Division, accessed May 29, 2017, http://www.mvd.usace.army.mil /About/Mississippi-River-Commission-MRC/Mississippi-River -Tributaries-Project-MR-T/Levee-Systems/.

24 Mark Schleifstein, "Louisiana's 2017 Coastal Master Plan Unanimously Approved by State Authority," *Times-Picayune,* April 19, 2017, accessed May 12, 2017, http://www.nola.com/environment/index.ssf/2017/04 /proposed_2017_coastal_master_p.html; "List of Louisiana Hurricanes (2000–Present)," *Wikipedia,* accessed May 12, 2017, https://en.wikipedia .org/wiki/List_of_Louisiana_hurricanes_(2000%E2%80%93present); "List of Oil Spills," *Wikipedia,* accessed May 12, 2017, https://en .wikipedia.org/wiki/List_of_oil_spills; Schleifstein, "As One Legislator Tears Up"; Emily Holden, "With a Master Plan and the Money, Can a State Unite to Restore Its Protective Wetlands?" *E&E News,* September 11, 2015, accessed May 12, 2017, https://www.eenews.net/stories /1060024532.

25 Schleifstein, "Louisiana's 2017 Coastal Master Plan"; "Pointe à la Hache, Louisiana," *Wikipedia*, accessed May 12, 2017, https://en.wikipedia.org /wiki/Pointe_%C3%A0_la_Hache,_Louisiana; Mark Schleifstein, "New Coastal Projects in Central, Western Louisiana Added to Master Plan," *Times-Picayune*, April 19, 2017, accessed May 12, 2017, http:// www.nola.com/environment/index.ssf/2017/04/new_projects_along _central_wes.html.

26 Quoted in Holden, "With a Master Plan."

27 Quoted in ibid.

28 Sara Sneath, "Nearly Three-Quarters of Louisiana Voters Believe Coastal Land Loss Will Affect Them: Poll," *Times-Picayune*, April 6, 2017, accessed May 28, 2017, http://www.nola.com/environment/index .ssf/2017/04/poll_97_percent_of_voters_stat.html; Bob Marshall, "Louisiana Keeps Voting to Drown: New Orleans Opinions," *Times-Picayune*, May 14, 2017, updated May 15, 2017, accessed May 28, 2017, http://www .nola.com/opinions/index.ssf/2017/05/louisiana_eroding_coast.html.

29 The account of the battle is from "Battle of Forts Jackson and St. Philip," *Wikipedia*, last modified March 28, 2017, accessed May 28, 2017, https://en .wikipedia.org/wiki/Battle_of_Forts_Jackson_and_St._Philip.

30 James L. Haley, *Sam Houston* (Norman: University of Oklahoma Press, 2004), 397.

31 Barry, *Rising Tide*, 67–92.

32 Baton Rouge Area Foundation, accessed May 28, 2017, http://www.braf.org/.

Chapter 7. The Fate of a Great City

1 Olga Bonfiglio, "The Fate of New Orleans Hangs in an Uncomfortable Balance with Mother Nature," *Resilience*, August 15, 2010, accessed May 29, 2017, http://www.resilience.org/stories/2010–08–15/fate-new-orleans -hangs-uncomfortable-balance-mother-nature.

2 Fred A. Bernstein, "Brad Pitt's Gifts to New Orleans," *New York Times*, November 25, 2009, accessed May 29, 2017, http://www.nytimes.com /2009/11/29/travel/29cultured.html; Tracy McVeigh, "The Dutch Solution to Floods: Live with Water, Don't Fight It," *Guardian*, February 15, 2014, accessed May 29, 2017, https://www.theguardian.com /environment/2014/feb/16/flooding-netherlands.

3 Zillow.com, accessed May 29, 2017, http://www.zillow.com/homes/for _sale/Lower-9th-Ward-New-Orleans-LA/pmf,pf_pt/269713_rid/30 .037594,-89.875889,29.858212,-90.173893_rect/11_zm/.

4 For a detailed view of the real estate situation, see Gary Rivlin, "Why New Orleans Black Residents Are Still Underwater after Katrina," *New York Times Magazine*, August 18, 2015, accessed May 29, 2017, https://www.nytimes.com/2015/08/23/magazine/why-new-orleans-black-residents-are-still-under-water-after-katrina.html?_r=0; also Greg Allen, "Ghosts of Katrina Still Haunt New Orleans' Shattered Lower 9th Ward," in the special series "Hurricane Katrina: 10 Years of Recovery and Reflection," NPR, August 3, 2015, accessed May 29, 2017, http://www.npr.org/2015/08/03/427844717/ghosts-of-katrina-still-haunt-new-orleans-shattered-lower-ninth-ward.

5 Throughout the chapter, quotations from Shirley Laska are from Laska, personal communication, January 13, 2012.

6 William R. Fredenburg et al., *Catastrophe in the Making: The Engineering of Katrina and the Disasters of Tomorrow* (Washington, DC: Island Press/Shearwater Books, 2009).

7 Ibid., 6–8.

8 Ibid., 10–11; Ari Kelman, *A River and Its City: The Nature of Landscape in New Orleans* (Berkeley: University of California Press, 2003); Ari Kelman, "City of Nature: New Orleans' Blessing; New Orleans' Curse," *Slate*, August 31, 2005, accessed May 29, 2017, http://www.slate.com/id/2125346; Peirce F. Lewis, *New Orleans: The Making of an Urban Landscape* (Santa Fe, NM: Center for American Places, 2003).

9 Fredenburg et al., *Catastrophe in the Making*; Flyer, "Facts for Features: Katrina Impact"; "Study: 1927 Mississippi Flood Would Cost up to $160 Billion," *Insurance Journal*, May 18, 2007, accessed May 28, 2017, http://www.insurancejournal.com/news/national/2007/05/18/79826.htm; William Faulkner, *Go Down, Moses* (New York: Random House, 1942), 346, quoted in Susan Scott Parrish, "Faulkner and the Outer Weather of 1927," *American Literary History* 24 (2012): 34–58, doi: https://doi.org/10.1093/alh/ajr052.

10 Fredenburg et al., *Catastrophe in the Making*, 118–134.

11 Ibid.

12 Gilbert F. White, "Human Adjustment to Floods," Department of Geography Research Paper no. 29, University of Chicago, 1945, 74.

13 Laska, personal communication.

14 Foster Creppel, personal communication, November 5, 2011.

15 Andrew W. Tweel and R. Eugene Turner, "Watershed Land Use and River Engineering Drive Wetland Formation and Loss in the Mississippi River Birdfoot Delta," *Limnology and Oceanography* 57 (2012): 18–28,

doi: 10.4319/lo.2012.57.1.0018; Robert H. Meade and John A. Moody, "Causes for the Decline of Suspended-Sediment Discharge in the Mississippi River System, 1940–2007," *Hydrological Progress* 24 (2010): 35–49, doi: 10.1002/hyp.7477; Glenn M. Suir et al., "Pictorial Account and Landscape Evolution of the Crevasses near Fort St. Phillip, Louisiana," *Mississippi River Geomorphology and Potamology Program, Report No. 2*, U.S. Army Corps of Engineers (2014): i–viii, 1–37, accessed May 29, 2017, https://pubs.er.usgs.gov/publication/70094484; Michael D. Blum and Harry H. Roberts, "Drowning of the Mississippi Delta Due to Insufficient Sediment Supply and Global Sea-Level Rise," *Nature Geoscience* 2 (2009): 488–491, doi: 10.1038/NGEO553.

16 Michael S. Kearney, J. C. Alexis Riter, and R. Eugene Turner, "Freshwater River Diversions for Marsh Restoration in Louisiana: Twenty-Six Years of Changing Vegetative Cover and Marsh Area," *Geophysical Research Letters* 38 (2011): L16405, doi: 10.1029/2011GL047847.

17 Laska, personal communication; Gilbert F. White, www.aag.org/cs /membership/tributes_memorials/sz/white_gilbert_f.

18 Ibid.

Chapter 8. Sea Level Rise Is Dangerous

1 Naomi Oreskes, "The Scientific Consensus on Climate Change," *Science* 306 (2004): 1686, doi: 10.1126/science.1103618; "Is Sea Level Rising?" Ocean Facts, National Ocean Service, National Oceanic and Atmospheric Administration, accessed February 13, 2017, http://oceanservice.noaa.gov/facts/sealevel.html; "Causes of Sea Level Rise: What the Science Tells Us," Union of Concerned Scientists, 2013, accessed March 4, 2017, http://www.ucsusa.org/global_warming/science _and_impacts/impacts/causes-of-sea-level-rise.html#.WKIBkLYrLGI; Svante Arrhenius, "On the Influence of Carbonic Acid in the Air upon the Temperature on the Ground," *Philosophical Magazine and Journal of Science* 5, no. 41 (1896): 237–276, accessed March 4, 2017, http://www .rsc.org/images/Arrhenius1896_tcm18-173546.pdf; "Svante Arrhenius," *Wikipedia*, accessed February 13, 2017, https://en.wikipedia.org/wiki /Svante_Arrhenius.

2 S. Jevrejeva et al., "Recent Global Sea Level Acceleration Started over 200 Years Ago," *Geophysical Research Letters* 35 (2008): L08715, doi: 10.1029/2008GL033611; "Historical Records May Underestimate Sea Level Rise," Vital Signs of the Planet, NASA Global Climate Change,

October 19, 2016, accessed March 4, 2017, http://climate.nasa.gov/news /2504/historical-records-may-underestimate-sea-level-rise/; Robert E. Kopp et al., "Temperature-Driven Global Sea-Level Variability in the Common Era," *Proceedings of the National Academy of Sciences USA* 113, no. 11 (2016): E1434–E1441, doi: 10.1073/pnas.1517056113; Stefan Rahmstorf, Grant Foster, and Anny Cazenave, "Comparing Climate Projections to Observations up to 2011," *Environmental Research Letters* 7 (November 27, 2012): 044035, doi: 10.1088/1748–9326/7/4/044035; Carbon Brief Staff, "Five Reasons Why the Speed of Arctic Sea Ice Loss Matters," *Carbon Brief/Science*, March 22, 2013, accessed March 4, 2017, https://www.carbonbrief.org/five-reasons-why-the-speed-of-arctic-sea -ice-loss-matters; Bobby Magill, "Arctic Methane Emissions 'Certain to Trigger Warming,'" *Climate Central*, May 1, 2014, accessed March 4, 2017, http://www.climatecentral.org/news/arctic-methane-emissions -certain-to-trigger-warming-17374; Rachael H. James et al., "Effects of Climate Change on Methane Emissions from Seafloor Sediments in the Arctic Ocean: A Review," *Limnology and Oceanography* 61 (November 2016): S283–S299, doi: 10.1002/lno.10307.

3 "Positive Feedback," *Wikipedia*, accessed February 13, 2017, https://en .wikipedia.org/wiki/Positive_feedback; Rahmstorf, Foster, and Caze- nave, "Comparing Climate Projections"; Barbara Neumann et al., "Future Coastal Population Growth and Exposure to Sea-Level Rise and Coastal Flooding—A Global Assessment," *PLoS One* 10 (March 11, 2015), http://dx.doi.org/10.1371/journal.pone.0118571; "What Percent- age of the American Population Lives near the Coast?" National Ocean Service, National Oceanic and Atmospheric Administration, accessed February 13, 2017, http://oceanservice.noaa.gov/facts/population.html; Stanley Chin, "What Percentage of the World's Population Lives Less Than a Meter above Sea-Level?" *Quora*, March 23, 2016, accessed March 4, 2017, https://www.quora.com/What-percentage-of-the-worlds -population-lives-less-than-a-meter-above-sea-level.

4 An RCP 8.5 (roughly "business as usual") high-emissions scenario corre- sponds to an increase in global temperature of 5.8 to 9.70 degrees Fahren- heit above 1850–1900 temperatures; see Malte Meinshausen et al., "The RCP Greenhouse Gas Concentrations and Their Extensions from 1765 to 2300," *Climatic Change* 109 (2011): 213–241, doi: 10.1007/s10584–011– 0156-z; J. A. Church et al., "Sea Level Change," in *Climate Change 2013: The Physical Science Basis. Contribution of Working Group I to the Fifth Assess- ment Report of the Intergovernmental Panel on Climate Change*, ed. T. F.

Stocker et al. (Cambridge, UK: Cambridge University Press, 2013), accessed March 4, 2017, http://www.climatechange2013.org/images/report /WG1AR5_Chapter13_FINAL.pdf; John Gregory, "Projections of Sea Level Rise," PowerPoint presented by the lead author of the IPCC Chapter 13 on sea level change to *Climate Change 2013*, accessed March 4, 2017, https://www.ipcc.ch/pdf/unfccc/cop19/3_gregory13sbsta.pdf.

5 "The World's Water," USGS Water School, U.S. Geological Survey, accessed February 14, 2017, https://water.usgs.gov/edu/earthwherewater .html; "State of the Cryosphere," National Snow and Ice Data Center, modified November 9, 2015, accessed March 4, 2017, https://nsidc.org /cryosphere/sotc/ice_sheets.html; Malcolm McMillan et al., "A High Resolution Record of Greenland Mass Balance," *Geophysical Research Letters* 43 (2016): 7002–7010, doi: 10.1002/2016GL069666; Robert M. De-Conto and David Pollard, "Contribution of Antarctica to Past and Future Sea-Level Rise," *Nature* 531 (2016): 591–597, doi: 10.1038/nature17145; Brenda Ekwurzel, "'Unstoppable' Destabilization of West Antarctic Ice Sheet May Have Been Crossed," *National Geographic/Ocean Views*, November 3, 2016, accessed March 4, 2017, http://voices.nationalgeographic .com/2016/11/03/unstoppable-destabilization-of-west-antarctic-ice-sheet -threshold-may-have-been-crossed/; Nicola Jones, "Abrupt Sea Level Rise Looms as Increasingly Realistic Threat," *Yale Environment* 360 (May 5, 2016), accessed March 4, 2017, http://e360.yale.edu/features /abrupt_sea_level_rise_realistic_greenland_antarctica; Justin Gillis, "Climate Model Predicts West Antarctic Ice Sheet Could Melt Rapidly," *New York Times*, March 30, 2016, accessed March 4, 2017, https://www .nytimes.com/2016/03/31/science/global-warming-antarctica-ice-sheet -sea-level-rise.html; Jugal L. Patel, "A Crack in an Antarctic Ice Shelf Grew 17 Miles in the Last Two Months," *New York Times*, February 7, 2017, accessed February 7, 2017, https://www.nytimes.com/interactive /2017/02/07/science/earth/antarctic-crack.html.

6 Martin J. Siegert, "Lakes beneath the Ice Sheet: The Occurrence, Analysis, and Future Exploration of Lake Vostok and Other Sub Antarctic Lakes," *Annual Review of Earth and Planetary Sciences* 33 (2005): 215–245, doi: 10.1146/annurev.earth.33.092203.122725; Martin J. Siegert, Neil Ross, and Anne M. LeBrocq, "Recent Advances in Understanding Antarctic Subglacial Lakes and Hydrology," *Philosophical Transactions of the Royal Society A* 374 (2016): 20140306, doi.org/10.1098/rsta.2014.0306.

7 Throughout the chapter, quotations by Rob Dunbar are from Dunbar, personal communications, May 5, 2017, and August 7, 2017.

8 "Mount Erebus," *Wikipedia*, accessed May 27, 2017, https://en.wikipedia .org/wiki/Mount_Erebus; "Brown Bluff," *Wikipedia*, accessed May 28, 2017, https://en.wikipedia.org/wiki/Brown_Bluff.

9 Dunbar, personal communication; H. Jay Zwally et al., "Surface Melt-Induced Acceleration of Greenland Ice-Sheet Flow," *Science* 297 (2002): 218–222, doi: 10.1126/science.1072708; Thomas L. Mote, "Greenland Surface Melt Trends, 1973–2007: Evidence of a Large Increase in 2007," *Geophysical Research Letters* 34 (2007), doi: 10.1029/2007GL031976.

10 Jonathan Kingslake et al., "Widespread Movement of Meltwater onto and across Antarctic Ice Shelves," *Nature* 544 (2017): 349–352, doi: 10.1038/nature22049.

11 Dunbar, personal communication; Jonathan L. Bamber et al., "Reassessment of the Potential Sea-Level Rise from a Collapse of the West Antarctic Ice Sheet," *Science* 324 (2009): 901–903, doi: 10.1126/science.1169335; Jugal K. Patel and Justin Gillis, "An Iceberg the Size of Delaware Just Broke Away from Antarctica," *New York Times*, updated July 12, 2017, accessed October 1, 2017, https://www.nytimes.com/interactive/2017 /06/09/climate/antarctica-rift-update.html?mcubz=1.

12 Patel and Gillis, "Iceberg the Size of Delaware."

13 Dunbar, personal communication; Robert M. DeConto and David Pollard, "Contribution of Antarctica to Past and Future Sea-Level Rise," *Nature* 531 (2016): 591–597, doi: 10.1038/nature17145.

14 Rob Dunbar, personal communication; T. Naish et al., "Obliquity-Paced Pliocene West Antarctic Ice Sheet Oscillations," *Nature* 458 (2009): 322–328, doi: 10.1038/nature07867; Carys P. Cook et al., "Dynamic Behavior of the East Antarctic Ice Sheet during Pliocene Warmth," *Nature Geoscience* 6 (2013), doi: 10.1038/NGEO1889.

15 S. Jevrejeva, A. Grinsted, and J. C. Moore, "Upper Limit for Sea Level Projections by 2100," *Environmental Research Letters* 9 (2014): 104008, doi: 10.1088/1748-9326/9/10/104008; Robert E. Kop et al., "Probabilistic 21st and 22nd Century Sea-Level Projections at a Global Network of Tide-Gauge Sites," *Earth's Future* 2 (2014): 383–406, doi: 10.1002/2014 EF000239; Matthias Mengel et al., "Future Sea Level Rise Constrained by Observations and Long-Term Commitment," *Proceedings of the National Academy of Sciences USA* 113, no. 10 (2016): 2597–2602, doi: 10.1073/pnas.1500515113; Gerard H. Roe and Marcia B. Baker, "Why Is Climate Sensitivity so Unpredictable?" *Science* 318 (2007): 629–632, doi: 10.1126/science.1144735; Trevir Nath, "Fat Tail Risk: What It Means and Why You Should Be Aware of It," *Nasdaq*, November 2, 2015,

accessed March 4, 2017, http://www.nasdaq.com/article/fat-tail-risk
-what-it-means-and-why-you-should-be-aware-of-it-cm537614.

16 Baden Copeland, Josh Keller, and Bill Marsh, "What Could Disappear,"
New York Times, modified April 24, 2016, accessed March 4, 2017, http://
www.nytimes.com/interactive/2012/11/24/opinion/sunday/what-could
-disappear.html.

17 Ibid.

18 Brenda Ekwurzel, "How Much Did Sea Levels Rise over the Past
50 Years? A Lot if You Live on the U.S. Gulf or East Coasts," *Union of
Concerned Scientists Blog*, May 2, 2014, accessed March 4, 2017, http://
blog.ucsusa.org/brenda-ekwurzel/how-much-did-sea-levels-rise-over
-the-past-50-years-a-lot-if-you-live-on-the-u-s-gulf-or-east-coasts-514.

19 Tal Ezer et al., "Gulf Stream's Induced Sea Level Rise and Variability
along the U.S. Mid-Atlantic Coast," *Journal of Geophysical Research* 118
(2013): 685–697, doi: 10.1002/jgrc.20091; Michael D. Lemonick, "East
Coast Faces Rising Seas from Slowing Gulf Stream," *Climate Central*,
February 12, 2013, accessed March 4, 2017, http://www.climatecentral
.org/news/east-coast-faces-rising-seas-from-slowing-gulf-stream-15587.

20 "Post-Glacial Rebound," *Wikipedia*, accessed February 14, 2017,
https://en.wikipedia.org/wiki/Post-glacial_rebound; Martin Ekman, *The
Changing Level of the Baltic Sea during 300 Years: A Clue to Understanding
the Earth* (Åland Islands, Finland: Summer Institute for Historical Geo-
physics, 2009), accessed March 4, 2017, https://www.google.com/search
?q=glacial+rebound+stockholm&rlz=1C5CHFA_enUS727US727&oq
=glacial+rebound+stockholm&aqs=chrome..69i57.10092j0j1&sourceid
=chrome&ie=UTF-8#q=post-glacial+rebound+stockholm+city+rising.

21 John Upton, "Sinking Atlantic Coastline Meets Rapidly Rising Seas,"
Climate Central, April 14, 2016, accessed March 4, 2017, http://www
.climatecentral.org/news/sinking-atlantic-coastline-meets-rapidly
-rising-seas-20247.

22 James P. M. Syvitski et al., "Sinking Deltas Due to Human Activities,"
Nature Geoscience 2 (2009): 681–686, doi: 10.1038/ngeo629; Rick Jervis,
"Oklahoma Earthquake Reignites Concerns That Fracking Wells May
Be the Cause," *USA Today*, November 7, 2016, accessed March 4, 2017,
http://www.usatoday.com/story/news/2016/11/07/oklahoma
-earthquake-fracking-well/93447830/; Matthew Phillips, "Why Oklahoma Can't Turn Off Its Earthquakes," *Bloomberg Businessweek*, November 8, 2016, accessed March 4, 2017, https://www.bloomberg.com/news
/articles/2016-11-08/why-oklahoma-can-t-turn-off-its-earthquakes;

William R. Freudenburg et al., *Catastrophe in the Making: The Engineering of Katrina and the Disasters of Tomorrow* (Washington, DC: Island Press, 2011), ISBN-10: 1610911636.

23 Syvitski et al., "Sinking Deltas Due to Human Activities."

24 "Mississippi River Delta," *Wikipedia*, accessed February 16, 2017, https://en.wikipedia.org/wiki/Mississippi_River_Delta; Michael D. Blum and Harry H. Roberts, "Drowning of the Mississippi Delta Due to Insufficient Sediment Supply and Global Sea-Level Rise," *Nature Geoscience* 2 (2009): 488–491, doi: 10.1038/NGEO553; Torbjörn E. Törnqvist et al., "Mississippi Delta Subsidence Primarily Caused by Compaction of Holocene Strata," *Nature Geoscience* 1 (2008): 173–176, doi: 10.1038/ngeo129; Jack Eggleston and Jason Pope, *Land Subsidence and Relative Sea-Level Rise in the Southern Chesapeake Bay Region*, U.S. Geological Survey Circular 1392 (2013), doi: 10.3133/cir1392.

25 "Coastal Erosion," *Wikipedia*, accessed February 16, 2017, https://en.wikipedia.org/wiki/Coastal_erosion.

26 Steve Chapple, "Into the Eye of a Hurricane," *Reader's Digest* (September 2009); "September 1948 Florida Hurricane," *Wikipedia*, accessed February 16, 2017, https://en.wikipedia.org/wiki/September_1948_Florida_hurricane; "Hurricane Carol," *Wikipedia*, accessed February 16, 2017, https://en.wikipedia.org/wiki/Hurricane_Carol; "Hurricane Allen," *Wikipedia*, accessed February 16, 2017, https://en.wikipedia.org/wiki/Hurricane_Allen; J. D. Woodley et al., "Hurricane Allen's Impact on Jamaican Coral Reefs," *Science* 214 (1981): 749–755, doi: 10.1126/science.214.4522.749.

27 "Tropical Cyclone Climatology," National Hurricane Center, National Oceanic and Atmospheric Administration, accessed February 16, 2017, http://www.nhc.noaa.gov/climo/; "Hurricanes Form over Tropical Oceans, Where Warm Water and Air Interact to Create These Storms," Ocean Explorer, National Oceanic and Atmospheric Administration, accessed February 16, 2017, http://oceanexplorer.noaa.gov/facts/hurricanes.html.

28 Staff writer, "Top 10 Most Expensive Disasters in U.S. History," *Trusted Choice*, Independent Insurance Agents, December 17, 2013, accessed March 4, 2017, https://www.trustedchoice.com/insurance-articles/weather-nature/most-expensive-disasters/; Denise Lu and Lazaro Gamio, "Here's Every Billion-Dollar Weather Disaster in the U.S. since 1980," *Washington Post*, August 25, 2015, modified January 10, 2017, accessed March 4, 2017, https://www.washingtonpost.com/graphics/national/billion-dollar-disasters/.

29 Kerry Emanuel, "Increasing Destructiveness of Tropical Cyclones over the Past 30 Years," *Nature* 436 (2005): 686–688, doi: 10.1038/nature03906; Michael E. Mann and Kerry A. Emanuel, "Atlantic Hurricane Trends Linked to Climate Change," *EOS: Earth and Space Science News* 87, no. 24 (2006): 233–241, doi: 10.1029/2006EO240001.

30 Ibid.; Thomas R. Knutson et al., "Tropical Cyclones and Climate Change," *Nature Geoscience* 3 (2010): 157–163, doi: 10.1038/ngeo779; James B. Elsner, James P. Kossin, and Thomas H. Jagger, "The Increasing Intensity of the Strongest Tropical Cyclones," *Nature* 455 (2008): 92–95, doi: 10.1038/nature07234; Nam-Young Kang and James B. Elsner, "Trade-Off between Intensity and Frequency of Global Tropical Cyclones," *Nature Climate Change* 5 (2015): 661–665, doi: 10.1038/NCLI MATE2646; Linda Lam, "New Evidence That Climate Change Is Altering Hurricane Season as You Know It," *The Weather Channel: Environment*, May 19, 2015, accessed March 4, 2017, https://weather.com /science/environment/news/climate-change-study-changing-intensity -number-hurricanes.

31 Chris Dolce, "Here's Why Hurricane Harvey Rapidly Intensified into a Category 4 Monster," *Weather Underground*, August 26, 2017, accessed October 1, 2017, https://www.wunderground.com/news/why-hurricane -harvey-intensified-rapidly.

32 Jonathan D. Woodruff, Jennifer L. Irish, and Suzana J. Camargo, "Coastal Flooding by Tropical Cyclones and Sea-Level Rise," *Nature* 504 (2013): 44–52, doi: 10.1038/nature12855; Nicholas K. Coch, "Unique Vulnerability of the New York–New Jersey Metropolitan Area to Hurricane Destruction," *Journal of Coastal Research* 31 (2014): 196–212, doi: 10.2112/JCOASTRES-D-13-00183.1; Jeroen C. J. H. Aerts et al., "Low-Probability Flood Risk Modeling for New York City," *Risk Analysis* 33 (2013): 772–788, doi: 10.1111/risa.12008; Christopher E. Schubert et al., *Analysis of Storm-Tide Impacts from Hurricane Sandy in New York*, U.S. Geological Survey Scientific Investigations Report (2015), 5036, accessed September 3, 2017, http://dx.doi.org/10.3133/sir20155036.

33 K. H. Jacob, V. Gornitz, and C. Rosenzweig, "Vulnerability of the New York City Metropolitan Area to Coastal Hazards, Including Sea-Level Rise: Inferences for Urban Coastal Risk Management and Adaptation Policies, in Managing Coastal Vulnerability," in *Managing Coastal Vulnerability: Global, Regional, Local*, ed. L. McFadden, R. J. Nicholls, and E. Penning-Roswell (Amsterdam: Elsevier, 2007), 61–88; N. Lin et al., "Risk Assessment of Hurricane Storm Surge for New York City,"

Journal of Geophysical Research Atmospheres 115, no. D18 (2010): doi: 10.1029/2009JD013630; Klaus Jacob et al., "Transportation," in *Responding to Climate Change in New York State: The ClimAID Integrated Assessment for Effective Climate Change Adaptation in New York State, Final Report No. 11–18*, ed. Cynthia Rosenzweig et al. (Albany, NY: NY-SERDA, 2011), 299–362; "Effects of Hurricane Irene in New York," *Wikipedia*, accessed February 16, 2017, https://en.wikipedia.org/wiki/Effects_of_Hurricane_Irene_in_New_York.

34 Andre Tartar, "Is Dr. Klaus H. Jacob the Cassandra of New York City Subway Flooding?" *New York Magazine, Daily Intelligencer/Hurricane Watch*, October 28, 2012, accessed March 4, 2017, http://nymag.com/daily/intelligencer/2012/10/climate-expert-warns-of-possible-subway-flooding.html; Bryan Walsh, "People Who Mattered in 2012: Klaus Jacob," *Time*, December 18, 2012, http://poy.time.com/2012/12/19/people-who-mattered-in-2012/slide/klaus-jacob/.

35 "Co-op City, Bronx," *Wikipedia*, accessed February 17, 2017, https://en.wikipedia.org/wiki/Co-op_City,_Bronx; Erica Pearson, "Troubles for Co-op City, and Middle Class Housing in NYC," *Gotham Gazette*, October 13, 2003, accessed March 4, 2017, http://www.gothamgazette.com/development/1992-troubles-for-co-op-city-and-middle-class-housing-in-nyc; Nina Wohl, "Co-op City: The Dream and the Reality," master's thesis, Columbia University, Graduate School of Arts and Sciences, American Studies Center for the Study of Ethnicity and Race, 2016, http://dx.doi.org/10.7916/D8MK6CZV.

36 William D. Nordhaus, "The Economics of Hurricanes and Implications of Global Warming," *Climate Change Economics* 1 (2010): 1–20, doi: 10.1142/S2010007810000054; William D. Nordhaus, *The Climate Casino: Uncertainty and Economics for a Warming World* (New Haven: Yale University Press, 2013), ISBN-10: 030018977X.

37 William D. Nordhaus, "Economic Policy in the Face of Severe Tail Events," *Journal of Public Economic Theory* 14 (2012): 197–219, doi: 10.1111/j.1467-9779.2011.01544.x; William D. Nordhaus, personal communication, October 29, 2014.

38 Nordhaus, personal communication.

Chapter 9. Retreat from the Coasts

1 Jeremy Jackson, "Sea Level Rise Is Dangerous," paper presented at an evening lecture at the U.S. Naval War College, Newport, Rhode Island,

February 4, 2015, accessed March 5, 2017, https://www.youtube.com
/watch?v=TAtCQ7REXAc.

2 Alon Harish, "New Law in North Carolina Bans Latest Scientific Predictions of Sea-Level Rise," *ABC News*, August 2, 2012, accessed March 5,
2017, http://abcnews.go.com/US/north-carolina-bans-latest-science
-rising-sea-level/story?id=16913782; Tristram Korten, "In Florida, Officials Ban Term 'Climate Change,'" *Miami Herald*, March 8, 2015, accessed March 5, 2017, http://www.miamiherald.com/news/state/florida
/article12983720.html; Leonard Pitts, Jr., "Gov. Scott: Do You Think if
We Ignore Climate Change, It Goes Away?" *Miami Herald*, March 15,
2016, accessed March 5, 2017, http://www.miamiherald.com/opinion/op
-ed/article66320997.html; Lisette Alvarez and Campbell Robertson,
"Cost of Flood Insurance Rises, Along with Worries," *New York Times*,
October 12, 2013, accessed March 5, 2017, http://www.nytimes.com
/2013/10/13/us/cost-of-flood-insurance-rises-along-with-worries.html;
Nicholas Nehamas, "Miami Real Estate Had a Really Bad Month in
July," *Miami Herald*, August 24, 2016, accessed March 5, 2017, http://
www.miamiherald.com/news/business/real-estate-news
/article97640357.html; Paul Owers, "South Florida Home Prices Rise
in October despite Sales Declines," *Sun Sentinel*, November 22, 2016,
accessed March 5, 2017, http://www.sun-sentinel.com/real-estate/news
/fl-home-sales-october-20161122-story.html.

3 Susan Hanson et al., "A Global Ranking of Port Cities with High Exposure to Climate Extremes," *Climatic Change* 104 (2011): 89–111, doi:
10.1007/s10584-010-9977-4.

4 Ibid.

5 Jeroen C. J. H. Aerts et al., "Evaluating Flood Resilience Strategies
for Coastal Megacities," *Science* 344 (2014): 473–475, doi: 10.1126/sci
ence.1248222.

6 Baden Copeland, Josh Keller, and Bill Marsh, "What Could Disappear,"
New York Times, modified April 24, 2016, accessed March 4, 2017, http://
www.nytimes.com/interactive/2012/11/24/opinion/sunday/what-could
-disappear.html; Tanvi Misra, "Mapping the Inundation of New York
City," *Atlantic/Citylab*, December 6, 2016, accessed March 5, 2017, http://
www.citylab.com/weather/2016/12/the-inundation-of-new-york-city
/509699/; Hanson et al., "Global Ranking"; *Rising Currents: Projects
for New York's Waterfront*, exhibit at the Museum of Modern Art,
March 24–October 11, 2010, accessed March 5, 2017, https://www.moma
.org/calendar/exhibitions/1028; *The Climate Museum—Creating a Hub for*

Climate Science, Art and Dialogue; A Beacon for Solutions, accessed February 20, 2017, http://climatemuseum.org/; Jeroen C. J. H. Aerts et al., "Evaluating Flood Resilience Strategies for Coastal Megacities," *Science* 344 (2014): 473–475, doi: 10.1126/science.1248222; Sarah Trefethen and Danielle Furfaro, "LaGuardia Rebuild Is Finally Underway," *New York Post*, June 1, 2016, accessed March 5, 2017, http://nypost.com/2016/06/01/la-guardia-rebuild-is-finally-underway/.

7 Elizabeth Kolbert, "Letter from Florida: The Siege of Miami," *New Yorker*, December 21 and 28, 2015, accessed March 5, 2017, http://www.newyorker.com/magazine/2015/12/21/the-siege-of-miami; Hanson et al., "Global Ranking"; "1926 Miami Hurricane," *Wikipedia*, accessed February 20, 2017, https://en.wikipedia.org/wiki/1926_Miami_hurricane; "The Hurricane of 1926," People and Events feature for the documentary "Mr. Miami Beach," *The American Experience*, Public Broadcasting System, 1998, accessed March 5, 2017, http://www.pbs.org/wgbh/amex/miami/peopleevents/pande07.html.

8 "1926 Miami Hurricane," *Wikipedia*; Frank Sessa, *Miami in 1926*, Florida International University Digital Collections, n.d., accessed March 5, 2017, http://digitalcollections.fiu.edu/tequesta/files/1956/56_1_02.pdf; "Miami Metropolitan Area," *Wikipedia*, accessed February 21, 2017, https://en.wikipedia.org/wiki/Miami_metropolitan_area.

9 "A Brief History of Real Estate on Key Biscayne, the Island Paradise," Key Life Realty, accessed March 5, 2017, http://www.keylife.com/history/; Natalie O'Neill, "The Quicksand Houses: A Miami Hood Is Sinking," *Miami New Times*, March 20, 2009, accessed March 5, 2017, http://www.miaminewtimes.com/news/the-quicksand-houses-a-miami-hood-is-sinking-6559051; Geniusofdespair, "Lennar's New Development's 'Prime Location,'" *Eye on Miami Blogspot*, June 24, 2012, accessed March 5, 2017, https://eyeonmiami.blogspot.com/2012/06/lennars-new-developments-prime-location.html; Kolbert, "Letter from Florida."

10 Pilkey is quoted in Mark Schrope, "Unarrested Development," *Nature Reports/Climate Change*, April 6, 2010, doi: 10.1038/climate.2010.27.

11 Anne Carrns, "How to Assess Private Flood Insurance," *New York Times*, September 7, 2016, accessed March 5, 2017, https://www.nytimes.com/2016/09/08/your-money/how-to-assess-private-flood-insurance.html; "Flood Insurance, National Flood Insurance Program," Insurance Information Institute, accessed February 20, 2017, http://www.iii.org/fact

-statistic/flood-insurance; Darryl Fears, "Rise in Government Insurance Rates to Mirror Rising Waters, Flood Debt," *Washington Post*, March 28, 2015, accessed March 5, 2017, https://www.washingtonpost.com/national/health-science/rise-in-government-insurance-rates-to-mirror-rising-waters-flood-debt/2015/03/28/8f9f17c6-d316-11e4-ab77-9646eea6a4c7_story.html?utm_term=.8ecf8ae1111b; Jeff Harrington, "Remember the Flood Insurance Scare of 2013? It's Creeping Back into Tampa Bay and Florida," *Tampa Bay Times*, August 5, 2016, accessed March 5, 2017, http://www.tampabay.com/news/business/banking/remember-the-flood-insurance-scare-of-2013-its-creeping-back-into-tampa/2288308; Lee Needleman, "Property Owners Beware, Flood Insurance Premiums on the Rise," *Miami Herald*, February 8, 2015, accessed March 5, 2017, http://www.miamiherald.com/news/business/biz-monday/article15559892.html.

12 "Citizens Property Insurance Corporation," *Wikipedia*, accessed February 21, 2017, https://en.wikipedia.org/wiki/Citizens_Property_Insurance_Corporation; Jim Turner, "State-Owned Citizens Insurance Numbers Might Rise," *News Chief*, December 9, 2016, accessed March 5, 2017, http://www.newschief.com/news/20161208/state-owned-citizens-insurance-numbers-might-rise.

13 Throughout the chapter, quotations by Robert Hartwig are from Hartwig, personal communication, September 11, 2014.

14 Brian McNoldy, "Water, Water, Everywhere: Sea Level Rise in Miami," *Rosenstiel School of Marine and Atmospheric Science Blogroll*, October 3, 2014, accessed March 5, 2017, http://www.rsmas.miami.edu/blog/2014/10/03/sea-level-rise-in-miami/; Pakalolo, "The Phenomenon That Can Not Be Spoken in Florida Continues as Salt Water Intrusion Moves Inland," *Daily Kos*, March 19, 2015, accessed March 5, 2017, http://www.dailykos.com/story/2015/3/19/1372031/-The-phenomenon-that-can-not-be-spoken-in-Florida-continues-as-salt-water-intrusion-moves-inland; Kyle Munzenrider, "Study: Miami Beach Will Experience 45 Floods a Year by 2030, 260 by 2045," *Miami New Times*, October 8, 2014, accessed March 5, 2017, http://www.miaminewtimes.com/news/study-miami-beach-will-experience-45-floods-a-year-by-2030-260-by-2045-6553631; Alexandra Witze, "Florida Forecasts Sinkhole Burden," *Nature* 504 (2013): 196–197, accessed March 7, 2017, http://www.nature.com/polopoly_fs/1.14337!/menu/main/topColumns/topLeftColumn/pdf/504196a.pdf.

15 D. Whittle and K. Lindeman, "Protecting Coastal Resources in Cuba: A Look at Current Laws and Institutions," in *Proceedings of the Coastal Society* (Newport, RI: Coastal Society 19th International Conference, 2004), 30–37, accessed March 5, 2017, http://citeseerx.ist.psu.edu /viewdoc/download?doi=10.1.1.459.8095&rep=rep1&type=pdf; Andrea Rodriguez, "Cuba Girds for Climate Change by Reclaiming Coasts," *Phys Org*, June 12, 2013, accessed March 5, 2016, https://phys.org/news /2013-06-cuba-girds-climate-reclaiming-coasts.html; Nelson Valdes, "Natural Disasters and Planning: The Cuban State and Popular Participation," *Counter Punch*, February 19, 2014, accessed March 5, 2017, http://www.counterpunch.org/2014/02/19/natural-disasters-and -planning-the-cuban-state-and-popular-participation/.

16 Lilah Raptopoulos, "'Are We Safe? Of Course Not': Climate Scientist's NYC Warning after Sandy," *Guardian*, November 5, 2014, accessed March 5, 2017, https://www.theguardian.com/us-news/2014/nov/05 /climate-scientist-klaus-jacob-warning-new-york-city-hurricane -sandy; "Thames Barrier," *Wikipedia*, accessed February 25, 2017, https://en.wikipedia.org/wiki/Thames_Barrier; Michael Kimmelman, "Going with the Flow," *New York Times*, February 13, 2013, http://www .nytimes.com/2013/02/17/arts/design/flood-control-in-the-netherlands -now-allows-sea-water-in.html.

17 *Are You Ready? Hurricane Season June 1–November 30: Your 2016 Guide to Hurricane Readiness*, Miami-Dade County government, accessed February 24, 2017, http://www.miamidade.gov/fire/library/hurricane/guide -to-hurricane-readiness.pdf; "Flood," Ready Colorado, accessed February 25, 2017, https://www.readycolorado.com/hazard/flood; "Flood Safety," National Weather Service, National Oceanic and Atmospheric Administration, accessed February 25, 2017, http://www.floodsafety .noaa.gov/; "Tsunami Facts and Preparedness," Office of Emergency Services, Government of San Diego County, accessed February 25, 2017, http://www.sandiegocounty.gov/oes/disaster_preparedness/oes _jl_tsunami.html; "Tsunami Signs," Caltrans, State of California, accessed February 25, 2017, http://www.dot.ca.gov/trafficops/tcd/tsunami .html; Valdes, "Natural Disasters and Planning."

18 Rafael Olmeda, "South Florida Returns to Normal, Spared from Hurricane Matthew's Worst," *Sun Sentinel*, October 7, 2016, accessed March 5, 2017, http://www.sun-sentinel.com/news/weather/hurricane/fl -hurricane-matthew-story.html; Jeremy Wallace, "Florida Recovers

from a Hurricane Matthew That Could Have Been Worse," *Miami Herald*, October 10, 2016, accessed March 5, 2017, http://www.miamiherald .com/news/weather/hurricane/article107380777.html.

19 Pitts, "Gov. Scott"; Greg Allen, "As Waters Rise, Miami Beach Builds Higher Streets and Political Willpower," National Public Radio, May 10, 2016, accessed March 5, 2017, http://www.npr.org/2016/05/10/476071206 /as-waters-rise-miami-beach-builds-higher-streets-and-political -willpower.

20 "New Orleans," *Wikipedia*, accessed February 25, 2017, https://en .wikipedia.org/wiki/New_Orleans; "New Orleans Topography," National Aeronautical and Space Administration, accessed February 25, 2017, https://www.nasa.gov/vision/earth/lookingatearth/h2005-neworleans -082905.html; Richard Campanella, "New Orleans Was Once above Sea Level, But Stormwater Drainage Has Caused It to Sink—With Deadly Consequences," *Times-Picayune*, January 28, 2016, accessed March 5, 2017, http://www.nola.com/homegarden/index.ssf/2015/02/shifting _doorframes_cracking_d.html; "New Study Maps Rate of New Orleans Sinking," NASA Jet Propulsion Laboratory, California Institute of Technology, May 16, 2016, accessed March 5, 2017, http://www.jpl.nasa.gov /news/news.php?feature=6513; Bob Marshall, "Losing Ground: Southeast Louisiana Is Disappearing Quickly," *Scientific American/Pro Publica*, August 28, 2014, accessed March 5, 2017, https://www.scientificamerican .com/article/losing-ground-southeast-louisiana-is-disappearing-quickly/; Roy Scranton, "When the Next Hurricane Hits Texas," *New York Times*, October 7, 2016, accessed March 5, 2017, https://www.nytimes.com/2016 /10/09/opinion/sunday/when-the-hurricane-hits-texas.html.

21 "Population Demographics for New Orleans, Louisiana in 2016 and 2017," *Suburban Stats: Current New Orleans, Louisiana, Population, Demographics and Stats in 2016, 2017*, accessed February 25, 2017, https:// suburbanstats.org/population/louisiana/how-many-people-live-in-new -orleans; Hanson et al., "Global Ranking"; John McQuaid, "'Cancer Alley': Myth or Fact?," *Times-Picayune*, May 24, 2000, accessed March 5, 2017, http://www.nola.com/politics/index.ssf/2000/05/cancer_alley_myth _or_fact.html; Rachel Cernasky, "Cancer Alley: Big Industry and Bigger Illness along Mississippi River," *Threehugger*, February 8, 2011, accessed March 5, 2017, http://www.treehugger.com/corporate-respon sibility/cancer-alley-big-industry-bigger-illness-along-mississippi-river .html.

22 Office of the Governor, Coastal Protection and Restoration Authority, accessed May 12, 2017, http://coastal.la.gov/; Alan Neuhauser, "Jindal Declares Climate Change a 'Trojan Horse,'" *U.S. News and World Report*, September 16, 2014, accessed March 5, 2017, https://www.usnews.com /news/articles/2014/09/16/louisianas-bobby-jindal-declares-climate -change-a-trojan-horse; Kit O'Connell, "Louisiana Floods Show Harm Caused by Climate Change Denial and Neglect of U.S. Infrastructure," *Mint Press News*, August 24, 2016, accessed March 5, 2017, http://www .mintpressnews.com/louisiana-floods-show-harm-caused-climate -change-denial-neglect-us-infrastructure/219731/; Megan Geuss, "Just One Candidate in Louisiana's Senate Runoff Embraces Climate Change Facts," *Ars Technica*, December 9, 2016, accessed March 5, 2017, https:// arstechnica.com/science/2016/12/just-one-candidate-in-louisianas -senate-runoff-embraces-climate-change-facts/; Paula Martinez, "Post-Katrina New Orleans Smaller, But Population Growth Rates Back on Track," National Public Radio, August 19, 2015, accessed March 5, 2017, http://www.npr.org/2015/08/19/429353601/post-katrina-new-orleans -smaller-but-population-growth-rates-back-on-track; Mark Schleif-stein, "Historic Lawsuit Seeks Billions in Damages from Oil, Gas, Pipeline Industries for Wetlands Losses," *Times-Picayune*, February 29, 2016, accessed March 5, 2017, http://www.nola.com/environment/index.ssf /2013/07/historic_east_bank_levee_autho.html; Tegan Wendland, "To Fight Coastal Damage, Louisiana Parishes Pushed to Sue Energy Industry," National Public Radio, January 23, 2017, http://www.npr.org/2017 /01/23/511216472/to-fight-coastal-damage-louisiana-parishes-pushed -to-sue-energy-industry; Mark Schleifstein, "Louisiana Coastal Restoration 50-Year Blueprint Released," *Times-Picayune*, January 12, 2012, accessed March 5, 2017, http://www.nola.com/environment/index.ssf /2012/01/louisiana_releases_50-year_blu.html; Bob Marshall, "2017 Coastal Master Plan Predicts Grimmer Future for Louisiana Coast as Worst-Case Scenario Becomes Best Case," *New Orleans Advocate*, January 3, 2017, accessed March 5, 2017, http://www.theadvocate.com/new _orleans/news/environment/article_5ac81e86-d1e7-11e6-9177 -1bbd55b599b7.html.

23 John McPhee, "The Control of Nature: Atchafalaya," *New Yorker*, February 23, 1987, accessed March 5, 2017, http://www.newyorker.com /magazine/1987/02/23/atchafalaya; William Sargent, "Letting Mississippi Run Its Natural Course Could Save New Orleans from Hurricanes," *Christian Science Monitor*, May 19, 2011, accessed March 5, 2017,

http://www.csmonitor.com/Commentary/Opinion/2011/0519/Letting
-Mississippi-run-its-natural-course-could-save-New-Orleans-from
-hurricanes.

24 Rob Dunbar, personal communication, May 5, 2017.

Chapter 10. Agriculture in Crisis

1 Economic Research Service, "Irrigation and Water Use: Overview,"
United States Department of Agriculture, last updated April 28, 2017,
accessed June 2, 2017, https://www.ers.usda.gov/topics/farm-practices
-management/irrigation-water-use/; Economic Research Service, "Irrigation and Water Use: Background," U.S. Department of Agriculture,
last updated April 28, 2017, accessed June 2, 2017, https://www.ers.usda
.gov/topics/farm-practices-management/irrigation-water-use
/background/; "Drought—Annual 2016," National Centers for Environmental Information, NOAA, accessed March 3, 2017, https://www.ncdc
.noaa.gov/sotc/drought/201613; Steve Ahillen, "Forest Fires Burn
119,000 Acres in 8 Southeastern States," *USA Today*, November 20,
2016, accessed June 2, 2017, https://www.usatoday.com/story/news
/nation-now/2016/11/20/forest-fires-burn-119000-acres-8-southeastern
-states/94169774/; Paul Gattis, "At Ground Zero of Alabama's Drought:
'It's an Agricultural Disaster,'" *AL.com*, October 23, 2016, accessed
March 3, 2017, http://www.al.com/news/huntsville/index.ssf/2016/10/at
_ground_zero_of_alabamas_dro.html; Jess Bidgood, "Scenes from
New England's Drought: Dry Wells, Dead Fish, and Ailing Farms,"
New York Times, September 26, 2016, accessed March 3, 2017, https://
www.nytimes.com/2016/09/27/us/scenes-from-new-englands-drought
-dry-wells-dead-fish-and-ailing-farms.html?_r=0.

2 Mark Gomez, "California Storms: This Rainy Season Now Ranks
2nd All Time in 122 Years of Records," *Mercury News*, April 7, 2017,
accessed May 31, 2017, http://www.mercurynews.com/2017/04/06
/california-storms-this-water-year-now-ranks-2nd-all-time-in-122
-years-of-records/; "Utah Is Nation's Fastest Growing State, Census Bureau Reports," U.S. Census Bureau Release no. CB16-214, accessed
March 3, 2017, http://www.census.gov/newsroom/press-releases/2016
/cb16-214.html; Phillip Reese, "Once a Boom State, California Sees a
Historic Period of Slow Population Growth," *Sacramento Bee*, April 4,
2016, accessed March 3, 2017, http://www.sacbee.com/site-services
/databases/article69054977.html; Marc Reisner, *Cadillac Desert: The*

American West and Its Disappearing Water (London: Penguin, 1993); Alexander E. M. Hess and Thomas C. Frohlich, "Seven States Running Out of Water," *USA Today,* May 25, 2014, accessed March 3, 2017, http://www.usatoday.com/story/money/business/2014/06/01/states-running-out-of-water/9506821/; *Chinatown* (1974 film), *Wikipedia,* modified on February 20, 2017, accessed March 3, 2017, https://en.wikipedia.org/wiki/Chinatown_(1974_film); Andrew Pulver, "*Chinatown*: The Best Film of All Time," *Guardian,* October 22, 2010, accessed March 3, 2017, https://www.theguardian.com/film/2010/oct/22/best-film-ever-chinatown-season.

3 The 2016 estimates for the metropolitan statistical areas are New York/Newark/Jersey City, NY/NJ, 20,153,634; Los Angeles/Long Beach/Anaheim, CA, 13,310,447; Chicago/Naperville/Elgin IL, 9,512,999; Dallas/Ft. Worth/Arlington, TX, 7,333,323; Houston/The Woodlands/Sugar Land, TX, 6,772,470, "List of Metropolitan Statistical Areas," *Wikipedia,* accessed May 29, 2017, https://en.wikipedia.org/wiki/List_of_Metropolitan_Statistical_Areas; "Drought—Annual 2016"; "List of United States Cities by Population," *Wikipedia,* accessed February 2, 2017, https://en.wikipedia.org/wiki/List_of_United_States_cities_by_population; Jordan Weissmann, "'Austin, Texas, Is Blowing Away Every Other Big City in Population Growth,' Moneybox," May 21, 2015, *Slate,* accessed March 3, 2017, http://www.slate.com/blogs/moneybox/2015/05/21/population_growth_in_u_s_cities_austin_is_blowing_away_the_competition.html.

4 Douglas A. McIntyre, "The Ten Biggest American Cities That Are Running Out of Water," *Wall Street Journal,* October 29, 2010, accessed February 2, 2017, http://247wallst.com/investing/2010/10/29/the-ten-great-american-cities-that-are-dying-of-thirst/print/; S. C. Gwynne, "The Last Drop," *Texas Monthly* (February 2008), accessed February 2, 2017, http://www.texasmonthly.com/articles/the-last-drop/.

5 "List of United States Cities by Population"; Matthew A. Winkler, "California Makes America's Economy Great," *Bloomberg View,* June 6, 2016, accessed March 3, 2017, https://www.bloomberg.com/view/articles/2016-06-06/california-makes-america-s-economy-great; Allison Vekshin, "California Overtakes France to Become Sixth-Largest Economy," *Bloomberg Politics,* June 14, 2016, accessed on February 2, 2017, https://www.bloomberg.com/politics/articles/2016-06-14/california-overtakes-france-to-become-sixth-largest-economy; Chris Nichols, "Does California Really Have the '6th Largest Economy on Planet Earth'?" *Politi-*

fact California, July 26, 2016, accessed March 4, 2017, http://www
.politifact.com/california/statements/2016/jul/26/kevin-de-leon/does
-california-really-have-sixth-largest-economy-/.

6 *California Agricultural Statistics Review, 2015–2016*, California Department of Food and Agriculture, accessed March 4, 2017, https://www.cdfa
.ca.gov/statistics/PDFs/2016Report.pdf; Brian Palmer, "The C-Free
Diet," *Slate*, July 10, 2013, accessed March 4, 2017, http://www.slate.com
/articles/health_and_science/explainer/2013/07/california_grows_all
_of_our_fruits_and_vegetables_what_would_we_eat_without.html;
Shannon Jones, "Top 5 Industries in California: Which Parts of
the Economy Are Strongest," *Newsmax*, February 25, 2015, accessed
March 4, 2017, http://www.newsmax.com/FastFeatures/industries-in
-california-strongest/2015/03/08/id/626901/; "California Is the World's
Sixth Largest Economy," *California's Economy*, accessed March 4, 2017,
http://www.lao.ca.gov/2000/calfacts/2000_calfacts_economy.pdf; "Rev-
enue Comparison of Apple, Google/Alphabet, and Microsoft from 2008
to 2015 (in Billion U.S. dollars)," *Statista: The Statistics Portal*, accessed
February 2, 2017, https://www.statista.com/statistics/234529/comparison
-of-apple-and-google-revenues/.

7 "California's Water Systems," *Maven's Notebook*, July 5, 2015, accessed
February 2, 2017; https://mavensnotebook.com/the-notebook-file
-cabinet/californias-water-systems/; "Where Does My Water Come
From?" *Water Education Foundation*, accessed February 2, 2017, http://
www.watereducation.org/where-does-my-water-come; George Kostyrko,
"Statewide Water Savings Nearly Reach 19 Percent in November; Most
of State Still Experiencing Drought Conditions," *California Drought*,
January 4, 2017, accessed February 2, 2017, http://drought.ca.gov/;
"Colorado River Compact," *Wikipedia*, accessed February 2, 2017,
https://en.wikipedia.org/wiki/Colorado_River_Compact; Ciara Dineen,
"Drought and California's Role in the Colorado River Compact," *Journal of Legislation* 42, no. 211 (2016), accessed March 4, 2017, http://
scholarship.law.nd.edu/jleg/vol42/iss2/4; Associated Press, "Western
Drought Watchers Eye Lake Mead Water Level," *Mercury News*, January 22, 2017, accessed March 4, 2017, http://www.mercurynews.com
/2017/01/20/western-drought-watchers-keep-wary-eye-on-lake-mead
-level/; see photographs in Eric Holthaus, "Lake Mead Before and
After the Epic Drought," *Slate*, July 25, 2014, accessed March 4, 2017,
http://www.slate.com/articles/technology/future_tense/2014/07/lake
_mead_before_and_after_colorado_river_basin_losing_water_at

_shocking.html; Gary Wockner, "Lake Powell: Going, Going, Gone?" *EcoWatch*, August 15, 2016, accessed March 4, 2017, http://www.ecowatch.com/lake-powell-drought-1974067342.html; David Gorn, "Desalination's Future in California Is Clouded by Cost and Controversy," *KQED Science*, October 31, 2016, accessed March 4, 2017, https://ww2.kqed.org/science/2016/10/31/desalination-why-tapping-sea-water-has-slowed-to-a-trickle-in-california/.

8 "Arizona's Water: Uses and Sources," Arizona Experience, accessed March 4, 2017, http://arizonaexperience.org/people/arizonas-water-uses-and-sources; Eric Holthaus, "Dry Heat: As Lake Mead Hits Record Lows and Water Shortages Loom, Arizona Prepares for the Worst," *Slate*, May 8, 2015, accessed March 4, 2017, http://www.slate.com/articles/health_and_science/science/2015/05/arizona_water_shortages_loom_the_state_prepares_for_rationing_as_lake_mead.html; Arizona Department of Water Resources, *Securing Arizona's Water Future*, accessed March 4, 2017, http://www.azwater.gov/AzDWR/PublicInformationOfficer/documents/supplydemand.pdf; Stephanie L. Castle et al., "Groundwater Depletion during Drought Threatens Future Water Security of the Colorado River Basin," *Geophysical Research Letters* 41 (2014): 5904–5911, doi: 10.1002/2014GL061055; Lindsay Ratcliff, "Massive Groundwater Losses Detected in the Colorado River Basin," *Yale Environment Review*, December 9, 2014, accessed March 4, 2017, http://environment.yale.edu/yer/article/massive-groundwater-losses-detected-in-the-colorado-river-basin#gsc.tab=0.

9 Robinson Meyer, "A Mega-Drought Is Coming to America's Southwest," *Atlantic*, October 11, 2016, https://www.theatlantic.com/science/archive/2016/10/megadroughts-arizona-new-mexico/503531/.

10 Noah S. Diffenbaugh et al., "Quantifying the Influence of Global Warming on Unprecedented Extreme Climate Events," *Proceedings of the National Academy of Sciences USA* 114 (2017): 4881–4886, doi: 10.1073/pnas.1618082114; Knutson et al., "Tropical Cyclones and Climate Change," *Nature Geoscience* 3 (2010): 157–163, doi: 10.1038/ngeo779; James B. Elsner, James P. Kossin, and Thomas H. Jagger, "The Increasing Intensity of the Strongest Tropical Cyclones," *Nature* 455 (2008): 92–95, doi: 10.1038/nature07234; Nam-Young Kang and James B. Elsner, "Trade-off between Intensity and Frequency of Global Tropical Cyclones," *Nature Climate Change* 5 (2015): 661–665, doi: 10.1038/NCLIMATE2646; Linda Lam, "New Evidence That Climate Change Is Altering Hurricane Season as You Know It," *Weather Channel: Environment*,

May 19, 2015, accessed March 4, 2017, https://weather.com/science /environment/news/climate-change-study-changing-intensity-number -hurricanes; Holly Evarts, "Increasing Tornado Outbreaks: Is Climate Change Responsible?" *Natural Disasters* blog, *State of the Planet*, Earth Institute/Columbia University, December 1, 2016, accessed June 2, 2017, http://blogs.ei.columbia.edu/2016/12/01/increasing-tornado-outbreaks -is-climate-change-responsible/; Michael K. Tippett, Chiara Lepore, and Joel E. Cohen, "More Tornadoes in the Most Extreme U.S. Tornado Outbreaks," *Science* 354 (2016): 1419–1423, doi: 10.1126/science .aah7393.

11 Benjamin I. Cook, Toby R. Ault, and Jason E. Smerdon, "Unprecedented 21st Century Drought Risk in the American Southwest and Central Plains," *Science Advances* 1 (February 1, 2015): e1400082, doi: 10.1126/sciadv.1400082.

12 Jeff Masters, "Ten Civilizations or Nations That Collapsed from Drought," *Wunderblog*, March 21, 2016, accessed March 4, 2017, https:// www.wunderground.com/blog/JeffMasters/ten-civilizations-or -nations-that-collapsed-from-drought; R. Kyle Bocinsky et al., "Exploration and Exploitation in the Macrohistory of the pre-Hispanic Pueblo Southwest," *Science Advances* (April 1, 2016): e1501532, doi: 10.1126/ sciadv.1501532; Nina von Uexkull et al., "Civil Conflict Sensitivity to Growing-Season Drought," *Proceedings of the National Academy USA* 113, no. 44 (2016): 12391–12396, www.pnas.org/cgi/doi/10.1073/pnas .1607542113; Benjamin I. Cook et al., "Spatiotemporal Drought Variability in the Mediterranean over the Last 900 Years," *Journal of Geophysical Research* 121 (2016): 2060–2074, doi: 10.1002/2015JD023929.

13 Cook, Ault, and Smerdon, "Unprecedented 21st Century Drought Risk."

14 J. Walsh et al., "Our Changing Climate," in *Climate Change Impacts in the United States: The Third National Climate Assessment*, ed. J. M. Melillo, Terese (T.C.) Richmond, and G. W. Yohe (Washington, DC: U.S. Global Change Research Program, 2014), 19–67, doi: 10.7930/J0KW5 CXT; "Extreme Precipitation Events Are on the Rise," *Climate Central*, May 6, 2014, accessed March 4, 2017, http://www.climatecentral.org /gallery/maps/extreme-precipitation-events-are-on-the-rise; "Climate Impacts in the Midwest," U.S. Environmental Protection Agency, accessed February 4, 2017, https://www.epa.gov/climate-impacts/climate -impacts-midwest; "Climate Impacts in the Northeast," U.S. Environmental Protection Agency, accessed February 4, 2017, https://www.epa .gov/climate-impacts/climate-impacts-northeast; "Climate Impacts in

the Southeast," U.S. Environmental Protection Agency, accessed February 4, 2017, https://www.epa.gov/climate-impacts/climate-impacts -southeast; "Snowmageddon," *Wikipedia*, accessed February 6, 2017, https://en.wikipedia.org/wiki/Snowmageddon; "Hurricane Sandy," *Wikipedia*, accessed February 6, 2017, https://en.wikipedia.org/wiki /Hurricane_Sandy; Angela Fritz, "Mid-Atlantic Coastline Flooded by Blizzard's Storm Surge. 'This Is Worse Than Sandy,'" *Washington Post*, January 24, 2016, accessed March 4, 2017, https://www.washingtonpost .com/news/capital-weather-gang/wp/2016/01/24/mid-atlantic -coastline-flooded-by-blizzards-storm-surge-this-is-worse-than-sandy /?utm_term=.02a408a598a3. It is noteworthy that much of this information is no longer directly available on the EPA website. Instead, one receives the following message: "This page is being updated. Thank you for your interest in this topic. We are currently updating our website to reflect EPA's priorities under the leadership of President Trump and Administrator Pruit. If you're looking for an archived version of this page, you can find it on the *January 19 snapshot*." Accessing the snapshot in May and June 2017, https://19january2017snapshot.epa.gov /climatechange_.html, did lead us to the original posting but with this proviso: "This is not the current EPA website . . . This website is historical material reflecting the EPA website as it existed on January 19, 2017. This website is no longer updated and its links to external websites and some internal pages may not work." We will be delighted to provide the original texts if they are permanently removed from the EPA website.

15 Christopher Doering, "Study: Iowa Lost 15 Million Tons of Soil to Erosion," *Des Moines Register*, August 4, 2014, accessed March 4, 2017, http://www.desmoinesregister.com/story/money/agriculture/2014/08 /05/environmental-working-group-says-iowa-lost-million-tons-soil -erosion/13609603/; Soren Rundquist and Craig Cox, "'Washout Revisited': Survey Finds Modest Progress Combatting Iowa Soil Erosion," Environmental Working Group, August 5, 2014, accessed March 4, 2017, http://www.ewg.org/research/washout-revisited; Donnelle Eller, "Erosion Estimated to Cost Iowa $1 Billion in Yield," *Des Moines Register*, May 3, 2014, http://www.desmoinesregister.com/story/money/agriculture /2014/05/03/erosion-estimated-cost-iowa-billion-yield/8682651/; Anna M. Michalak et al., "Record-Setting Algal Bloom in Lake Erie Caused by Agricultural and Meteorological Trends Consistent with Expected Future Conditions," *Proceedings of the National Academy of Sciences USA* 110, no. 16 (2013): 6448–6452, doi: 10.1073/pnas.1216006110;

Mahdi M. Al-Kaisi et al., "Drought Impact on Crop Production and the Soil Environment: 2012 Experiences from Iowa," *Journal of Soil and Water Conservation* 68 (2013): 19A–24A, doi: 10.2489/jswc.68.1.19A.

16 Wolfram Schlenker and Michael J. Roberts, "Nonlinear Temperature Effects Indicate Severe Damages to U.S. Crop Yields under Climate Change," *Proceedings of the National Academy of Sciences USA* 106 (2009): 15594–15598, doi: 10.1073/pnas.0906865106.

17 Elizabeth Weise, "Some Crops Migrate North with Warmer Temperatures," *USA Today*, September 17, 2013, accessed March 4, 2017, http://www.usatoday.com/story/news/nation/2013/09/17/climate-change-agriculture-crops/2784561/; Alan Bjerga, "Canada's Cornbelt Attracts the Hot Money," *Bloomberg*, November 8, 2012, accessed March 4, 2017, https://www.bloomberg.com/news/articles/2012-11-08/canadas-corn-belt-attracts-the-hot-money; Alan Bjerga, "Canada's Climate Warms to Corn as Grain Belt Shifts North," *Bloomberg*, April 15, 2014, accessed March 4, 2017, https://www.bloomberg.com/news/articles/2014-04-15/canada-s-climate-warms-to-corn-as-grain-belt-shifts-north; S. Asseng et al., "Rising Temperatures Reduce Global Wheat Production," *Nature Climate Change* 5 (2014): 143–147, doi: 10.1038/nclimate2470; L. C. Bliss et al., "Vegetation Regions," *Historica Canada*, last modified March 4, 2015, accessed March 4, 2017, http://www.thecanadianencyclopedia.ca/en/article/vegetation-regions/.

18 Tim Radford, "Study Finds Plant Growth Surges as CO_2 Levels Rise," *Climate Central*, June 9, 2013, accessed March 4, 2017, http://www.climatecentral.org/news/study-finds-plant-growth-surges-as-co2-levels-rise-16094; "C_3 Carbon Fixation," *Wikipedia*, accessed February 6, 2017, https://en.wikipedia.org/wiki/C3_carbon_fixation; "C_4 Carbon Fixation," *Wikipedia*, accessed February 6, 2017, https://en.wikipedia.org/wiki/C4_carbon_fixation; Matt Ridley, "Two Rival Kinds of Plants and Their Future," *MattRidley Online Blog*, July 7, 2012, accessed March 4, 2017, http://www.rationaloptimist.com/blog/two-rival-kinds-of-plants-and-their-future/; Lewis H. Ziska, "The Impact of Elevated CO_2 on Yield Loss from a C_3 and C_4 Weed in Field-Grown Soybean," *Global Change Biology* 6 (2010): 899–905, doi: 10.1046/j.1365-2486.2000.00364.x; Andrew D. B. Leakey, "Rising Atmospheric Carbon Dioxide Concentration and the Future of C_4 Crops for Food and Fuel," *Proceedings of the Royal Society B* 276 (2009): 2333–2343, doi: 10.1098/rspb.2008.1517; David B. Lobell et al., "Greater Sensitivity to Drought Accompanies Maize Yield Increase in the U.S. Midwest,"

Science 344 (2014): 516–519, doi: 10.1126/science.1251423; Daniel Urban et al., "Projected Temperature Changes Indicate Significant Increase in Interannual Variability of U.S. Maize Yields," *Climatic Change* 112 (2012): 525–533, doi: 10.1007/s10584–012–0428–2.

19 Timothy Griffin et al., "Regional Self-Reliance of the Northeast Food System," *Renewable Agriculture and Food Systems* 30, no. 4 (2015), accessed March 4, 2017, https://doi.org/10.1017/S1742170514000027; "Global Climate Change Impacts in the United States, 2009 Report: Agriculture," U.S. Global Change Research Program, accessed March 4, 2017, https://nca2009.globalchange.gov/agriculture/index.html; S. Grund and E. Walberg, *Climate Change Adaptation for Agriculture in New England* (Plymouth, MA: Manomet Center for Conservation Sciences, 2013), accessed March 4, 2017, https://www.manomet.org/sites/default/files/publications _and_tools/Agriculture_fact_sheet%205-13.pdf; Doug Fraser, "Climate Change: Heat May Drive Cranberry Industry North," *Cape Cod Times*, October 27, 2015, accessed March 4, 2017, http://www.capecodtimes .com/news/20151026/climate-change-heat-may-drive-cranberry -industry-north; Sarah Brown, "Global Warming Pushes Maple Trees, Syrup to the Brink," *National Geographic/The Plate*, December 2, 2015, accessed March 4, 2017, http://theplate.nationalgeographic.com /2015/12/02/global-warming-pushes-maple-trees-syrup-to-the-brink/; "Climate Change Impacts: Climate Impacts in the Northeast," U.S. Environmental Protection Agency, accessed February 12, 2017, https:// www.epa.gov/climate-impacts/climate-impacts-northeast; Michael Casey and Patrick White, "Spread by Trade and Climate, Bugs Butcher America's Forests," *PhysOrg*, December 7, 2016, accessed March 4, 2017, https:// phys.org/news/2016-12-climate-bugs-butcher-america-forests.html.

20 *Obesity Update*, Organization for Economic Cooperation and Development (OECD), June 2014, accessed March 4, 2017, http://www.oecd.org /els/health-systems/Obesity-Update-2014.pdf; Cynthia L. Ogden et al., *Prevalence of Obesity among Adults and Youth: United States, 2011–2014*, National Center for Health Statics (NCHS) Data Brief no. 219, November 2015, accessed March 4, 2017, https://www.cdc.gov/nchs/data/databriefs /db219.pdf; Ross A. Hammond and Ruth Levine, "The Economic Impact of Obesity in the United States," *Diabetes, Metabolic Syndrome and Obesity: Targets and Therapy* 3 (2010): 285–295, doi: 10.2147/DMSOTT.S7384; "Economic Costs of Obesity," National League of Cities, accessed February 12, 2017, http://www.healthycommunitieshealthyfuture.org/learn-the -facts/economic-costs-of-obesity/.

21 "List of Countries by Life Expectancy," *Wikipedia*, accessed February 12, 2017, https://en.wikipedia.org/wiki/List_of_countries_by_life _expectancy; Mauricio Avendano and Ichiro Kawachi, "Why Do Americans Have Shorter Life Expectancy and Worse Health Than People in Other High-Income Countries?" *Annual Review of Public Health* 35 (2014): 307–325, doi: 10.1146/annurev-publhealth-032013-182411.

22 Joeri Rogelj, Malte Meinshausen, and Reto Knutti, "Global Warming under Old and New Scenarios Using IPCC Climate Sensitivity Range Estimates," *Nature Climate Change* 2 (2012): 248–253, doi: 10.1007/s10584-012-0428-2; Joeri Rogelj et al., "Differences between Carbon Budget Estimates Unraveled," *Nature Climate Change* 8 (2016): 245–252, doi: 10.1038/nclimate2868.

Chapter 11. Reinventing Agriculture

1 "The Climate Is What You Expect; the Weather Is What You Get," Quote Investigator, June 24, 2012, accessed March 4, 2017, http://quoteinvestigator.com/2012/06/24/climate-vs-weather/; Associated Press, "California Caps Off Year with More Rain, Snow," *Mercury News*, January 1, 2017, accessed March 4, 2017, http://www.mercurynews.com /2016/12/31/californias-wet-december-drawing-to-a-close-with-more -rain-2/.

2 "Billion-Dollar Weather and Climate Disasters: Summary Stats," NOAA, National Centers for Environmental Information, accessed March 4, 2017, https://www.ncdc.noaa.gov/billions/summary-stats.

3 Marc Reisner, *Cadillac Desert: The American West and Its Disappearing Water* (London: Penguin, 1993).

4 "Billion-Dollar Weather and Climate Disasters"; Mark Bittman, "A Simple Fix for Farming," *New York Times*, October 19, 2012, accessed March 4, 2017, https://opinionator.blogs.nytimes.com/2012/10/19/a -simple-fix-for-food/; Rodale Institute, "Farming Systems Trial: Celebrating 30 Years," accessed May 31, 2017, http://rodaleinstitute.org /assets/FSTbookletFINAL.pdf; Paul McMahon, *Feeding Frenzy* (London: Profile Books, 2013).

5 Michael M. Bell, "Did New England Go Downhill?" *Geographical Review* 79 (1989): 450–466, accessed March 4, 2017, http://www.jstor.org /stable/215118; John W. Wysong, Mary G. Leigh, and Pradeep Ganguly, "The Economic Viability of Commercial Fresh Vegetable Production in the Northeastern United States," *Journal of the Northeastern Agricultural*

Economics Council 13 (1984): 65–72 (succeeding title: *Northeastern Journal of Agricultural and Resource Economics*), accessed March 4, 2017, http://purl.umn.edu/159498; "Vegetables and Pulses," U.S. Department of Agriculture, Economic Research Service, last modified January 9, 2017, accessed March 4, 2017, https://www.ers.usda.gov/topics/crops/vegetables-pulses/; Sara D. Short, *Characteristics and Production Costs of U.S. Dairy Operations*, U.S. Department of Agriculture, Statistical Bulletin no. 974–6 (2004), accessed March 4, 2017, https://www.ers.usda.gov/webdocs/publications/sb9746/28499_sb974-6_1_.pdf; Farm Credit East, *Northeast Agriculture: The Overlooked Economic Engine*, 2012, accessed March 4, 2017, http://www.zwickcenter.uconn.edu/documents/farmcrediteconomicimpact.pdf.

6 "Governor's Drought Declaration," California Department of Water Resources, accessed March 4, 2017, http://www.water.ca.gov/water conditions/declaration.cfm; "Statewide Water Savings Nearly Reach 19 Percent in November; Most of State Still Experiencing Drought Conditions," *California Drought*, January 4, 2017, accessed March 4, 2017, http://drought.ca.gov/; Mark Gold, "California Is Backsliding on Water Conservation. L.A. Can't and Won't Follow Suit," *Los Angeles Times*, September 12, 2016, accessed March 4, 2017, http://www.latimes.com/opinion/livable-city/la-ol-california-water-conservation-20160912-snap-story.html; Phillip Reese and Ryan Sabalow, "California Backing Away from Strict Water-Saving Standards," *Sacramento Bee*, May 9, 2016, http://www.sacbee.com/news/state/california/water-and-drought/article76553182.html.

7 *2016 Annual Report, Climate Action Plan*, City of San Diego, accessed March 4, 2017, https://www.sandiego.gov/sites/default/files/2016_annual_report_climate_action_plan.pdf; Debra Prinzing, "More California Gardeners Go Native," *Los Angeles Times*, October 16, 2014, accessed March 4, 2017, http://www.latimes.com/home/la-hm-native-gardening-20141016-story.html; Kurtis Alexander, "Water Guzzlers Face More Fines, Public Shaming under Environmental Law," *San Francisco Chronicle*, December 31, 2016, accessed March 4, 2017, http://www.sfchronicle.com/bayarea/article/Water-guzzlers-face-more-fines-public-shaming-10828966.php; Tom Philpott, "There's a Place That's Nearly Perfect for Growing Food. It's Not California," *Mother Jones*, April 20, 2015, accessed March 4, 2017, http://www.motherjones.com/tom-philpott/2015/04/decalifornify-cotton-vegetables-fruit-south.

8 "Water Supply and Sanitation in Israel," *Wikipedia,* accessed February 8, 2017, https://en.wikipedia.org/wiki/Water_supply_and_sanitation_in _Israel; Joel Greenberg, "Israel No Longer Worried about Its Water Supply, Thanks to Desalination Plants," *McClatchy DC Bureau,* March 20, 2014, accessed March 4, 2017, http://www.mcclatchydc.com /news/nation-world/national/article24765472.html; "Existing and Proposed Seawater Desalination Plants in California," Pacific Institute, modified May 2016, accessed March 4, 2017, http://pacinst.org/publication /key-issues-in-seawater-desalination-proposed-facilities/.

9 "Claude 'Bud' Lewis Carlsbad Desalination Plant," Carlsbad Desalination Project, accessed February 7, 2017, http://carlsbaddesal.com/; Bradley J. Fikes, "State's Biggest Desal Plant to Open: What It Means," *San Diego Union-Tribune,* December 13, 2015, accessed March 4, 2017, http:// www.sandiegouniontribune.com/news/environment/sdut-poseidon -water-desalination-carlsbad-opening-2015dec13-htmlstory.html; *Seawater Desalination: The Claude "Bud" Lewis Desalination Plant and Related Facilities,* San Diego County Water Authority, accessed February 7, 2017, http://www.sdcwa.org/sites/default/files/desal-carlsbad-fs-single.pdf.

10 "Reverse Osmosis," *Wikipedia,* accessed February 7, 2017, https://en .wikipedia.org/wiki/Reverse_osmosis; Heather Cooley and Kristina Donnelly, *Key Issues in Seawater Desalination in California: Proposed Seawater Desalination Facilities,* Pacific Institute, July 2012, accessed March 4, 2017, http://pacinst.org/app/uploads/2014/04/desalination-facilities.pdf.

11 "By the Numbers: Desalination Energy Usage," Carlsbad Desalination Project, accessed March 4, 2017, http://carlsbaddesal.com/Websites /carlsbaddesal/images/Fact_Sheets/Energy_Infographic_2015.pdf; "San Diego Gas and Electric Generation Fact Sheet," San Diego Gas and Electric, September 29, 2014, accessed February 7, 2017, http://www .sdge.com/sites/default/files/newsroom/factsheets/SDG%26E%20 Electric%20Generation%20Fact%20Sheet_2.pdf.

12 Rob Dunbar, personal communication, May 5, 2017; Pacific Institute/ NRDC, *The Untapped Potential of California's Water Supply,* Pacific Institute, June 2014, accessed March 4, 2017, http://pacinst.org/app/uploads /2014/06/ca-water-capstone.pdf; Matt Weiser, "New California Dam Proposed to Combat Climate Change Concerns," *News Deeply/Water Deeply,* January 9, 2017, accessed March 4, 2017, https://www.newsdeeply .com/water/articles/2017/01/09/new-california-dam-proposed-to -combat-climate-change-concerns; Benjamin Gross, "Los Angeles Has

a Plan to Start Fixing Up the Whole LA River," *Curbed Los Angeles,* January 20, 2015, accessed March 4, 2017, http://la.curbed.com/2015/1/20/10001016/los-angeles-has-a-plan-to-start-fixing-up-the-whole-la-river; Peter H. Gleick, "California Must Capture Water, Not Waste It," *Los Angeles Times,* November 17, 2015, accessed March 4, 2017, http://www.latimes.com/opinion/op-ed/la-oe-1117-gleick-groundwater-strategies-20151117-story.html; Jacques Leslie, "Los Angeles, City of Water," *New York Times,* December 6, 2014, accessed March 4, 2017, https://www.nytimes.com/2014/12/07/opinion/sunday/los-angeles-city-of-water.html; Matt Weiser, "California: Catching Up with the Irrigation World," *News Deeply/Water Deeply,* June 22, 2016, accessed March 4, 2017, https://www.newsdeeply.com/water/community/2016/06/22/california-catching-up-with-the-irrigation-world; Pamela Nagappan, "California Farmers Innovate to Fight Drought," *News Deeply/Water Deeply,* May 13, 2016, accessed March 4, 2017, https://www.newsdeeply.com/water/articles/2016/05/13/california-farmers-innovate-to-fight-drought.

13 Heather Hacking, "Oroville Dam: Who Pays for Spillway Repairs?" *Mercury News,* May 30, 2017, accessed May 30, 2017, http://www.mercurynews.com/2017/05/30/oroville-dam-who-pays-for-spillway-repairs/.

14 Todd Woody, "Holy Cow! Crops That Use Even More Water Than Almonds," *Take Part,* May 11, 2015, accessed March 4, 2017, http://www.takepart.com/article/2015/05/11/cows-not-almonds-are-biggest-water-users/; Mark Bittman, "Not All Industrial Food Is Evil," *New York Times,* August 17, 2013, accessed March 4, 2017, https://opinionator.blogs.nytimes.com/2013/08/17/not-all-industrial-food-is-evil/.

15 M. A. White et al., "Extreme Heat Reduces and Shifts United States Premium Wine Production in the 21st Century," *Proceedings of the National Academy of Sciences USA* 103 (2006): 11217–11222, doi: 10.1073/pnas.0603230103; Union of Concerned Scientists, "Napa Valley, CA, USA," *Climate Hot Map: Global Warming Effects Around the World,* accessed June 3, 2017, http://www.climatehotmap.org/global-warming-locations/napa-valley-ca-usa.html; Michelle Renée Mozell and Liz Thach, "The Impact of Climate Change on the Global Wine Industry: Challenges and Solutions," *Wine Economics and Policy* 3 (2014): 81–89, http://doi.org/10.1016/j.wep.2014.08.001.

16 Mark Bittman, "A Simple Fix for Farming," *New York Times,* October 19, 2012, accessed March 4, 2017, https://opinionator.blogs.nytimes.com

/2012/10/19/a-simple-fix-for-food/; Matt Liebman and Lisa Schulte, "Enhancing Agroecosystem Performance and Resilience through Increased Diversification of Landscapes and Cropping Systems," *Elementa: Science of the Anthropocene* 3 (2015): 41, accessed March 4, 2017, http://doi.org/10.12952/journal.elementa.000041.

17 Adam S. Davis et al., "Increasing Cropping System Diversity Balances Productivity, Profitability and Environmental Health," *PLoS One* 7 (2012): e47149, doi: 10.1371/journal.pone.0047149; Brandon Keim, "Big, Smart and Green: A Revolutionary Vision for Modern Farming," *Wired*, October 19, 2012, accessed March 4, 2017, https://www.wired.com/2012/10/big-smart-green-farming/.

18 Meghann E. Jarchow et al., "Trade-off among Agronomic, Energetic, and Environmental Performance Characteristics of Corn and Prairie Bioenergy Cropping Systems," *Global Change Biology Bioenergy* 7 (2015): 57–71, doi: 10.1111/gcbb.12096; Matt Liebman and Lisa A. Schulte, "Enhancing Agroecosystem Performance and Resilience through Increased Diversification of Landscapes and Cropping Systems," *Elementa* 3 (2015), doi: 10.12952/journal.elementa.000041.

19 Jarchow et al., "Trade-off."

20 Ibid. For the beneficial ecosystem effects of incorporating prairie into agricultural practices, see Matt Liebman et al., "Using Biodiversity to Link Agricultural Productivity with Environmental Quality: Results from Three Field Experiments in Iowa," *Renewable Agriculture and Food Systems* 28, no. 2 (2013): 115–128; Matthew J. Helmers et al., "Sediment Removal by Prairie Filter Strips in Row-Cropped Ephemeral Watersheds," *Journal of Environmental Quality* 41 (2012): 1531–1539; Xiaobo Zhou et al., "Nutrient Removal by Prairie Filter Strips in Agricultural Landscapes," *Journal of Soil and Water Conservation* 69 (2014): 54–64.

21 Marc I. Stutter, Wim J. Chardon, and Brian Kronvang, "Riparian Buffer Strips as a Multifunctional Tool in Agricultural Landscapes: Introduction," *Journal of Environmental Quality* 41 (2012): 297–303, accessed March 4, 2017, doi: 10.2134/jeq2011.0439; Leibman and Schulte, "Enhancing Agroecosystem Performance."

22 Paul M. Mayer et al., "Meta-Analysis of Nitrogen Removal in Riparian Buffers," *Journal of Environmental Quality* 36 (2006): 1172–1180, accessed March 4, 2017, doi: 10.2134/jeq2006.0462; X. Zhou et al., "Nutrient Removal by Prairie Filter Strips in Agricultural Landscapes," *Journal of Soil and Water Conservation* 69 (2014): 54–64, accessed

March 4, 2017, doi: 10.2489/jswc.69.1.54; "Billion-Dollar Weather and Climate Disasters."

23 Sarah M. Hirsh et al., "Diversifying Agricultural Catchments by Incorporating Tallgrass Prairie Buffer Strips," *Ecological Restoration* 31 (2013): 201–211, accessed March 4, 2017, doi: 10.3368/er.31.2.201; Bruce A. Robertson et al., "Perennial Biomass Feedstocks Enhance Avian Diversity," *Global Change Biology Bioenergy* 3 (2011): 235–246, accessed March 4, 2017, doi: 10.1111/j.1757–1707.2010.01080.x.

24 Ben P. Werling et al., "Perennial Grasslands Enhance Biodiversity and Multiple Ecosystem Services in Bioenergy Landscapes," *Proceedings of the National Academy of Sciences USA* 111 (2014): 1652–1657, accessed March 4, 2017, www.pnas.org/cgi/doi/10.1073/pnas.1309492111; Mary A. Gardiner et al., "Implications of Three Biofuel Crops for Beneficial Arthropods in Agricultural Landscapes," *BioEnergy Research* 3 (2010): 6–19, accessed March 4, 2017, doi: 10.1007/s12155–009–9065–7.

25 Daniel J. Conley et al., "Controlling Eutrophication: Nitrogen and Phosphorus," *Science* 323 (2009): 1014–1015, doi: 10.1126/science.1167755; Allen G. Good and Perrin H. Beatty, "Fertilizing Nature: A Tragedy of Excess in the Commons," *PLOS Biology*, August 16, 2011, http://doi.org /10.1371/journal.pbio.1001124; "Controlled Drainage," *Soil Science Online!*, North Carolina State University, accessed June 2, 2017, http:// www.soil.ncsu.edu/publications/BMPs/drainage.html; "Reducing Nutrient Loss: Science Shows What Works," Iowa State University Extension and Outreach, accessed June 2, 2017, https://www.cals.iastate .edu/sites/default/files/misc/183758/sp435.pdf; Thomas Stokes, "Removal of Nitrogen and Phosphorus from Tile Drainage Water," master's thesis, Iowa State University, 2016, https://masters.agron.iastate .edu/files/stokesthomas-finalccproj.pdf; "Taxes on Pesticides and Chemical Fertilizers," U.N. Development Program, accessed June 2, 2017, http://www.undp.org/content/sdfinance/en/home/solutions/taxes -pesticides-chemicalfertilizers.html; Christian Schwägerl, "With Too Much of a Good Thing, Europe Tackles Excess Nitrogen," *Yale Environment 360* (2015), accessed June 2, 2017, http://e360.yale.edu/features /with_too_much_of_a_good_thing_europe_tackles_excess_nitrogen.

26 *Understanding Federal Subsidies for the Biofuels and Biomass Industries*, Taxpayers for Common Sense, September 2015, accessed March 4, 2017, http://www.taxpayer.net/images/uploads/articles/biofuel-report-sept -15.pdf; Mark A. McMinimy and Kelsi Bracmort, "Renewable Fuel Standard (RFS): Overview and Issues," *Congressional Research Service*, No-

vember 22, 2013, accessed March 4, 2017, http://www.lankford.senate
.gov/imo/media/doc/1st%20Renewable%20Fuel%20Standard%20
Overview%20and%20Issues.pdf; Jerry Shenk, "Federal Government's
Ethanol Mandate Hoses You at the Gas Pump," *PennLive*, August 22,
2013, accessed March 4, 2017, http://www.pennlive.com/opinion/2013
/08/jerry_shenk_ethanol_is_corporate_welfare_harms_environment
.html; Amory Lovins, *Reinventing Fire: Bold Business Solutions for the New
Energy Era* (Hartford, VT: Chelsea Green, 2011).

27 Lovins, *Reinventing Fire*; Irene Kwan, "Planes, Trains, and Automobiles:
Counting Carbon," *International Council on Clean Transportation Blog*,
September 19, 2013, accessed March 4, 2017, http://www.theicct.org
/blogs/staff/planes-trains-and-automobiles-counting-carbon; Man in
Seat Sixty-One, "CO2 Emissions and Climate Change: Trains versus
Planes," accessed February 9, 2017, http://www.seat61.com/CO2flights
.htm; Sami Grover, "Is Train Travel Really the Greenest of the Green?"
Treehugger, January 23, 2009, accessed March 4, 2017, http://www
.treehugger.com/cars/trains-vs-planes-is-rail-always-the-low-carbon
-option.html; "The Northeast Corridor: Critical Infrastructure for the
Northeast Economy," Amtrak, accessed February 9, 2017, https://nec
.amtrak.com/sites/default/files/NEC%20Fact%20Sheet%20Win-
ter%202014_2.pdf; Ron Nixon, "Frustrations of Air Travel Push Pas-
sengers to Amtrak," *New York Times*, August 15, 2012, accessed March 4,
2017, http://www.nytimes.com/2012/08/16/business/hassles-of-air-travel
-push-passengers-to-amtrak.html.

28 "Regional Census Data by Year," U.S. Census Bureau, accessed March 4,
2017, https://www.census.gov/popclock/data_tables.php?component
=growth; Richard Florida, "The Dozen Regional Powerhouses Driving the U.S. Economy," *Atlantic/Citylab*, March 12, 2014, accessed
March 4, 2017, http://www.citylab.com/work/2014/03/dozen-regional
-powerhouses-driving-us-economy/8575/; Beth Kowitt, "Special Report: The War on Big Food," *Fortune*, May 21, 2015, accessed March 4,
2017, http://fortune.com/2015/05/21/the-war-on-big-food/; Merrill
Douglas, "The New Food Entrepreneurs," Cornell Enterprise, S. C.
Johnson College of Business, Cornell University, July 2016, accessed
March 4, 2017, https://www.johnson.cornell.edu/CornellEnterprise
/Article/ArticleId/44518/The-New-Food-Entrepreneurs; David R. Foster et al., *Wildlands and Woodlands: A Vision for the New England Land-
scape* (Cambridge, MA: Harvard University Press, 2010), accessed
March 4, 2017, http://www.wildlandsandwoodlands.org/sites/default

/files/Wildlands%20and%20Woodlands%20New%20England.pdf; Brian Donahue et al., *A New England Food Vision* (Durham, NH: Food Solutions New England, Sustainability Institute, University of New Hampshire, 2014), http://www.foodsolutionsne.org/sites/default/files /LowResNEFV_0.pdf; "Northeast Megalopolis," *Wikipedia*, accessed February 12, 2017, https://en.wikipedia.org/wiki/Northeast_megalopolis.

29 Farm Credit East, *Northeast Agriculture.*

30 Timothy Griffin et al., "Regional Self-Reliance of the Northeast Food System," *Renewable Agriculture and Food Systems* 30, no. 4 (2015), accessed March 4, 2017, https://doi.org/10.1017/S1742170514000027; "Global Climate Change Impacts in the United States, 2009 Report: Agriculture," U.S. Global Change Research Program, accessed March 4, 2017, https:// nca2009.globalchange.gov/agriculture/index.html; S. Grund and E. Walberg, *Climate Change Adaptation for Agriculture in New England* (Plymouth, MA: Manomet Center for Conservation Sciences, 2013), accessed March 4, 2017, https://www.manomet.org/sites/default/files /publications_and_tools/Agriculture_fact_sheet%205-13.pdf; Doug Fraser, "Climate Change: Heat May Drive Cranberry Industry North," *Cape Cod Times*, October 27, 2015, accessed March 4, 2017, http://www .capecodtimes.com/news/20151026/climate-change-heat-may-drive -cranberry-industry-north; Sarah Brown, "Global Warming Pushes Maple Trees, Syrup to the Brink," *National Geographic/The Plate*, December 2, 2015, accessed March 4, 2017, http://theplate.nationalgeographic .com/2015/12/02/global-warming-pushes-maple-trees-syrup-to-the -brink/; "Climate Change Impacts: Climate Impacts in the Northeast," U.S. Environmental Protection Agency, accessed February 12, 2017, https://www.epa.gov/climate-impacts/climate-impacts-northeast; Michael Casey and Patrick White, "Spread by Trade and Climate, Bugs Butcher America's Forests," *PhysOrg*, December 7, 2016, accessed March 4, 2017, https://phys.org/news/2016-12-climate-bugs-butcher -america-forests.html.

31 Catherine Greene et al., "Growing Organic Demand Provides High-Value Opportunities for Many Types of Producers," *Amber Waves*, U.S. Department of Agriculture, Economic Research Service, February 6, 2017, accessed June 2, 2017, https://www.ers.usda.gov/amber-waves/2017 /januaryfebruary/growing-organic-demand-provides-high-value -opportunities-for-many-types-of-producers/; Monica Orrigo, "Top Ten Food Industry Trends for 2016," *Handshake: B2B Commerce Digital*, January 4, 2016, accessed June 2, 2017, https://www.handshake.com/blog

/food-industry-trends/; David W. Wolfe, *Climate Change Impacts on Northeast Agriculture: Overview*, factsheet, Cornell University, Department of Horticulture, 2006, accessed March 4, 2017, https://www.uvm.edu/vtvegandberry/ClimateChange/ClimateChangeImpactsNortheastAgriculture.pdf; Leon Neyfakh, "How New England Could Become Farmville Again," *Boston Globe*, November 21, 2014, accessed March 4, 2017, https://www.bostonglobe.com/ideas/2014/11/21/how-new-england-could-become-farmville-again/mwD67IVd69wF9iLJFzyuuO/story.html.

32 "The Farming Systems Trial: Celebrating 30 Years," Rodale Institute, accessed June 2, 2017, http://rodaleinstitute.org/assets/FSTbooklet FINAL.pdf.

33 "Reaping the Benefits: Science and the Sustainable Intensification of Global Agriculture," Royal Society, October 2009, accessed June 3, 2017, https://royalsociety.org/~/media/Royal_Society_Content/policy /publications/2009/4294967719.pdf; Paul McMahon, "The Investment Case for Ecological Farming," SLM Partners white paper, January 2016, http://slmpartners.com/wp-content/uploads/2016/01/SLM-Partners -Investment-case-for-ecological-farming.pdf.

34 Frederick, edited by Jason C. Chavis, "Advantages of Industrial Agriculture," blog at *Bright Hub*, November 13, 2013, accessed June 2, 2017, http://www.brighthub.com/environment/science-environmental /articles/73606.aspx; Jayson Lusk, "Why Industrial Farms Are Good for the Environment," *New York Times*, September 23, 2016, accessed June 2, 2017, https://www.nytimes.com/2016/09/25/opinion/sunday /why-industrial-farms-are-good-for-the-environment.html?_r=0; Nina Federoff and Nancy Marie Brown, *Mendel in the Kitchen: A Scientist's View of Genetically Modified Foods* (Washington, DC: Joseph Henry Press, 2004); Ken Kimmell, "Industrial Agriculture: The Outdated, Unsustainable System that Dominates U.S. Food Production," Union of Concerned Scientists, May 24, 2017, accessed June 2, 2017, http://www .ucsusa.org/our-work/food-agriculture/our-failing-food-system /industrial-agriculture#.WTHCJhPytRo; McMahon, "Investment Case for Ecological Farming."

Epilogue

Epigraph. James D. Newton, *Uncommon Friends: Life with Thomas Edison, Henry Ford, Harvey Firestone, Alexis Carrel, & Charles Lindbergh* (San Diego:

Harcourt Brace Jovanovich, 1987), 31; for an intriguing discussion of the accuracy of this quotation, since Newton's book was published some fifty years after he sat with Edison, see Quote Investigator, accessed March 15, 2017, http://quoteinvestigator.com/2015/08/09/solar/.

1 Eric M. Conway and Naomi Oreskes, *Merchants of Doubt* (London: Bloomsbury Press, 2010); "Svante Arrhenius," *Wikipedia*, accessed June 13, 2017, https://en.wikipedia.org/wiki/Svante_Arrhenius; Justin Gillis, "A Scientist, His Work, and a Climate Reckoning," *New York Times*, December 21, 2010, accessed June 13, 2017, http://www.nytimes.com/2010/12/22/science/earth/22carbon.html; Danny Hakin, "Doubts about the Promised Bounty of Genetically Modified Crops," *New York Times*, October 29, 2016, accessed March 20, 2017, https://www.nytimes.com/2016/10/30/business/gmo-promise-falls-short.html; Andrew Kimbrell, "The Big Lie: Monsanto and the *New York Times*," *Huffington Post Blog*, December 15, 2016, updated December 16, 2016, accessed March 20, 2017, http://www.huffingtonpost.com/andrew-kimbrell/the-big-lie-monsanto-and-_b_13654204.html; Danny Hakim, "Monsanto Weed Killer Roundup Faces New Doubts on Safety in Unsealed Documents," *New York Times*, March 14, 2017, accessed March 15, 2017, https://www.nytimes.com/2017/03/14/business/monsanto-roundup-safety-lawsuit.html; Dan Charles, "Emails Reveal Monsanto's Tactics to Defend Glyphosate against Cancer Fears," *The Salt*, National Public Radio, March 15, 2017, accessed June 13, 2017, http://www.npr.org/sections/thesalt/2017/03/15/520250505/emails-reveal-monsantos-tactics-to-defend-glyphosate-against-cancer-fears; Upton Sinclair, *I, Candidate for Governor: And How I Got Licked* (1935; Berkeley: University of California Press, 1994).

2 Coral Davenport, "Trump Plans to Begin E.P.A. Rollback with Order on Clean Water," *New York Times*, February 28, 2017, accessed June 13, 2017, https://www.nytimes.com/2017/02/28/us/politics/trump-epa-clean-water-climate-change.html; Nathan Rott and Merrit Kennedy, "Trump Takes Aim at a Centerpiece of Obama's Environmental Legacy," *The Two-Way: Breaking News from NPR*, March 28, 2017, accessed June 12, 2017, http://www.npr.org/sections/thetwo-way/2017/03/28/519003733/trump-takes-aim-at-a-centerpiece-of-obamas-environmental-legacy; Michael R. Bloomberg and Jerry Brown, "The U.S. Is Tackling Global Warming Even If Trump Isn't," *New York Times*, November 14, 2017, downloaded November 14, 2017, https://www.nytimes.com/2017/11/14/opinion/global-warming-paris-climate-agreement.html; Michael Birn-

baum and Greg Jaffe, "Frustrated Foreign Leaders Bypass Washington in Search of Blue-State Allies," *Washington Post*, November 18, 2017, downloaded November 18, 2017, https://www.washingtonpost.com/world /national-security/frustrated-foreign-leaders-bypass-washington-in -search-of-blue-state-allies/2017/11/17/3ad10e80-cbab-11e7-8321-481fd 63f174d_story.html?utm_term=.f2a258fe6cf0; Mark Muro and Sifan Liu, "Another Clinton-Trump Divide: High Output America vs Low-Output America," *Brookings: The Avenue*, November 29, 2016, downloaded November 25, 2017, https://www.brookings.edu/blog/the-avenue/2016/11 /29/another-clinton-trump-divide-high-output-america-vs-low-output -america/; https://www.usatoday.com/story/weather/2017/11/07/u-s-now -only-country-not-part-paris-climate-agreement-after-syria-signs /839909001/.

3 Scott Clement and Brady Dennis, "Post-ABC poll: Nearly 6 in 10 Oppose Trump Scrapping Paris Agreement," *Washington Post*, June 5, 2017, accessed June 5, 2017, https://www.washingtonpost.com/news/energy -environment/wp/2017/06/05/post-abc-poll-nearly-6-in-10-oppose -trump-scrapping-paris-agreement/?hpid=hp_hp-top-table-main _climatepoll-308pm%3Ahomepage%2Fstory&utm_term= .a9ced206144f; Lisa Hymas, "Here Are the Oil and Coal Companies, Fortune 500 Corporations, and Republicans Who Want to Stay in the Paris Agreement," *Media Matters for America*, May 31, 2017, accessed June 5, 2017, https://www.mediamatters.org/blog/2017/05/31/Here-are -the-oil-and-coal-companies-Fortune-500-corporations-and -Republicans-who-want-to-s/216719; Stephen Pacala and Robert Socolow, "Stabilization Wedges: Solving the Climate Problem for the Next 50 Years with Current Technologies," *Science* 305 (2004): 968–972, doi: 10.1126/science.1100103; Paul Hawken, ed., *Drawdown: The Most Comprehensive Plan Ever Proposed to Reverse Global Warming* (London: Penguin Books, 2017); Amory B. Lovins, *Soft Energy Paths: Towards a Durable Peace* (New York: Harper & Row, 1979); "Amory B. Lovins," Rocky Mountain Institute, accessed March 20, 2017, http://www.rmi.org /Amory+B.+Lovins; Amory B. Lovins, *Reinventing Fire* (White River Junction, VT: Chelsea Green, 2011); Mark Z. Jacobson et al., "100% Clean and Renewable Wind, Water, and Sunlight (WWS) All-Sector Energy Roadmaps for the 50 United States," *Energy and Environmental Science* 8 (2015): 2093–2117, doi: 10.1039/C5EE01283J; Mark Z. Jacobson, Mark A. Delucchi, Mary A. Cameron, and Bethany A. Frew, "Low-Cost Solution to the Grid Reliability Problem with 100% Penetration of Wind, Water,

and Solar for All Purposes," *Proceedings of the National Academy of Sciences USA* 112 (2015): 15060–15065, doi: 10.1073/pnas.1510028112.

4 Pacala and Socolow, "Stabilization Wedges"; Robert Socolow, "Wedges Reaffirmed," *Bulletin of the Atomic Scientists,* September 27, 2011, accessed March 20, 2017, http://thebulletin.org/wedges-reaffirmed; Chris Mooney, *The Republican War on Science* (New York: Basic Books, 2005), 352. Socolow also co-authored two reports for the National Academies: National Academy of Sciences, National Academy of Engineering, and National Research Council of the National Academies, *America's Energy Future: Technology and Transformation* (Washington, DC: National Academies Press, 2010); and National Research Council, *America's Climate Choices* (Washington, DC: National Academies Press, 2011); http://science.sciencemag.org/content/305/5686/968.full.

5 Lovins, *Soft Energy Paths;* "Amory B. Lovins," Rocky Mountain Institute; "Amory Lovins," *Wikipedia,* last updated March 5, 2017, accessed March 20, 2017, https://en.wikipedia.org/wiki/Amory_Lovins.

6 Lovins, *Reinventing Fire.*

7 Amory B. Lovins, "Four Trends Driving Profitable Climate Protection," *Forbes,* September 21, 2015, accessed March 20, 2017, https://www.forbes.com/sites/amorylovins/2015/09/21/four-trends-driving-profitable-climate-protection/#7ec88fe73f4d; "Fossil Fuel Subsidies: Overview," *Oil Change International,* accessed June 8, 2017, http://priceofoil.org/fossil-fuel-subsidies/; David Coady et al., "How Large Are Global Fossil Fuel Subsidies?" *World Development* 91 (2017): 11–27, doi.org/10.1016/j.worlddev.2016.10.004; Damian Carrington and Harry Davies, "US Taxpayers Subsidizing World's Biggest Fossil Fuel Companies," *Guardian,* May 12, 2015, accessed June 6, 2017, https://www.theguardian.com/environment/2015/may/12/us-taxpayers-subsidising-worlds-biggest-fossil-fuel-companies; Eduardo Porter, "Do Oil Companies Really Need $4 Billion Per Year of Taxpayers' Money?" *New York Times,* August 5, 2016, accessed June 8, 2017, https://www.nytimes.com/2016/08/06/upshot/do-oil-companies-really-need-4-billion-per-year-of-taxpayers-money.html?_r=0.

8 Center for Responsible Politics, "Influence and Lobbying, Oil and Gas: Summary," Open Secrets.org., accessed June 8, 2017, https://www.opensecrets.org/industries/indus.php?cycle=2016&ind=E01.

9 International Energy Agency, "World Energy Investment 2017," July 11, 2017, accessed September 26, 2017, https://www.iea.org/publications/wei2017/; Fred Pearce, "Investment in Renewables Was Double That

of Coal and Gas in 2015," *New Scientist*, March 24, 2016, accessed October 1, 2017, https://www.newscientist.com/article/2082235-investment-in-renewables-was-double-that-of-coal-and-gas-in-2015/; Lovins, "Four Trends," emphasis in the original.

10 Michael Liebreich, "Renewable Energy Excl[uding] Large Hydro, Proportion of Power Generation, 2006–16," presentation, Bloomberg New Energy Finance Summit, April 25, 2017, accessed June 8, 2017, https://data.bloomberglp.com/bnef/sites/14/2017/04/2017-04-25-Michael-Liebreich-BNEFSummit-Keynote.pdf; Lovins, "Four Trends."

11 Lovins, "Four Trends."

12 Jacobson et al., "100% Clean and Renewable"; see also Bjorn Carey, "Stanford Engineers Develop State-by-State Plan to Convert U.S. to 100% Clean, Renewable Energy by 2050," *Stanford News Service*, June 8, 2015, accessed March 20, 2017, http://news.stanford.edu/pr/2015/pr-50states-renewable-energy-060815.html.

13 Carey, "Stanford Engineers"; Tom Dart and Oliver Milman, "'The Wild West of Wind': Republicans Push Texas as Unlikely Green Energy Leader," *Guardian*, February 20, 2017, accessed March 20, 2017, https://www.theguardian.com/us-news/2017/feb/20/texas-wind-energy-green-turbines-repbublicans-environment; Ari Shapiro and Matt Ozug, "Wind Energy Takes Flight in the Heart of Texas Oil Country," National Public Radio, March 8, 2017, accessed March 20, 2017, http://www.npr.org/2017/03/08/518988840/wind-energy-takes-flight-in-the-heart-of-texas-oil-country; as quoted in Marc Gunther, "The 100% Clean Energy Economy Is Closer Than You Think," *Sierra*, April 6, 2016, accessed March 20, 2017, http://www.sierraclub.org/sierra/2016-3-may-june/feature/100-percent-clean-energy-economy-closer-you-think; Justin Gillis and Nadja Popovich, "In Trump Country, Renewable Energy Is Thriving," *New York Times*, June 6, 2017, accessed June 8, 2017, https://www.nytimes.com/2017/06/06/climate/renewable-energy-push-is-strongest-in-the-reddest-states.html?mwrsm=Email.

14 Eduardo Porter, "Fisticuffs over the Route to a Clean Energy Future," *New York Times*, June 20, 2017, accessed July 29, 2017, https://www.nytimes.com/2017/06/20/business/energy-environment/renewable-energy-national-academy-matt-jacobson.html; Christopher T. M. Clack et al., "Evaluation of a Proposal for Reliable Low-Cost Power with 100% Wind, Water, and Solar," *Proceedings of the National Academy of Sciences USA* 114 (2017): 6722–6727, www.pnas.org/cgi/doi/10.1073/pnas.1610381114; Mark Z. Jacobson, Mark A. Delucchi, Mary A. Cameron,

and Bethany A. Frew, "The United States Can Keep the Grid Stable at Low Cost with 100% Clean, Renewable Energy in All Sectors Despite Inaccurate Claims," *Proceedings of the National Academy of Sciences USA* 114 (2017): E5021–5023, www.pnas.org/cgi/doi:10.1073/pnas.1708069114; NOAA, "Rapid Affordable Energy Transformation Possible," January 25, 2016, accessed September 26, 2017, http://www.noaanews.noaa.gov /stories2016/012516-rapid-affordable-energy-transformation-possible .html; Alexander E. MacDonald et al., "Future Cost-Competitive Electricity Systems and Their Impact on US CO_2 Emissions," *Nature Climate Change* 6 (published online January 25, 2016): 526–531, doi: 10.1038/ NCLIMATE2921.

15 Jack Ewing, "Volvo, Betting on Electric, Moves to Phase Out Conventional Engines," *New York Times*, July 5, 2017, accessed September 4, 2017, https://www.nytimes.com/2017/07/05/business/energy-environment /volvo-hybrid-electric-car.html; Brad Plummer, "When Will Electric Cars Go Mainstream? It May Be Closer Than You Think," *New York Times*, July 8, 2017, accessed September 4, 2017, https://www.nytimes .com/2017/07/08/climate/electric-cars-batteries.html.

16 April Baumgarten, "Wind Energy Sector Sees Massive Expansion in North Dakota," *Bismarck Tribune*, February 12, 2017, accessed October 1, 2017, http://bismarcktribune.com/news/state-and-regional/wind-energy -sector-sees-massive-expansion-in-north-dakota/article_b29c4e50-f8be -5399-ad46-5ee3811c8965.html; "GE Lead Wind Technician Hourly Pay," Glassdoor.com, accessed September 26, 2017, https://www .glassdoor.com/Hourly-Pay/GE-Lead-Wind-Technician-Hourly-Pay -E277_D_KO3,23.htm.

17 Liebreich, "Renewable Energy."

18 Oliver Milman, "San Diego Republican Mayor Pushes Plan to Run on 100% Renewable Energy," *Guardian*, April 26, 2016, accessed June 8, 2017, https://www.theguardian.com/us-news/2016/apr/26/san-diego-mayor -renewable-energy-plan-kevin-faulconer.

19 Ibid.

20 "SDG&E to Install Thousands of Electric Vehicle Charging Stations," *SDGE Newsroom*, January 28, 2016, accessed March 20, 2017, http://www .sdge.com/newsroom/press-releases/2016-01-28/sdge-install-thousands -electric-vehicle-charging-stations; leading Tesla cities would appear to be Norway's Oslo, with San Francisco, Los Angeles, San Diego, and Dallas in the United States, Reddit post May 16, 2015, accessed March 20, 2017, https://www.reddit.com/r/teslamotors/comments/367ea5/which

_cities_have_the_most_amount_of_teslas/; with Tesla Model S sales-leading states, in order, California, Florida, and Texas, according to Mark Kane, "California Leads Nation in Tesla Model S Sales, but Which Other States Are in Top 10?" *InsideEVs*, September 14, 2015, accessed March 20, 2017, http://insideevs.com/california-leads-nation-in-tesla-model-s-sales-but-which-other-states-are-in-top-10/.

21 For Cindy Hooven at the UC San Diego Extension, see http://extension.ucsd.edu/about/index.cfm?vAction=instructorBio&personid=212768, accessed March 20, 2017; for the Hackathon, see Tiffany Fox, "Smart City Hackathon Connects the Dots on Behalf of Climate," *UC San Diego News Center*, June 13, 2016, accessed March 20, 2017, http://ucsdnews.ucsd.edu/pressrelease/smartcity_hackathon_connects_the_data_dots_on_behalf_of_climate.

22 Cindy Hooven, personal communication, June 2016; Daniel Obodovski, personal communication, June 2016; Daniel Kellmereit and Daniel Obodovski, *The Silent Intelligence: The Internet of Things* (Warwick, NY: DND Ventures LLC, 2013).

23 Hooven, personal communication.

24 Ibid.

25 Adam Chandler, "Why Americans Lead the World in Food Waste," *Atlantic*, July 15, 2016, accessed June 8, 2017, https://www.theatlantic.com/business/archive/2016/07/american-food-waste/491513/; Suzanne Goldenberg, "Half of All U.S. Food Produce Is Thrown Away, New Research Suggests," *Guardian*, July 13, 2016, accessed June 8, 2017, https://www.theguardian.com/environment/2016/jul/13/us-food-waste-ugly-fruit-vegetables-perfect?CMP=share_btn_tw.

26 Emily S. Rueb, "How New York Is Turning Food Waste into Compost and Gas," *New York Times*, June 2, 2017, accessed June 8, 2017, https://www.nytimes.com/2017/06/02/nyregion/compost-organic-recycling-new-york-city.html; Tyler Drown, "How Many Apartment Buildings Are There in NYC (Exclude Condo/Co-ops)?" *Quora*, January 12, 2015, accessed June 8, 2017, https://www.quora.com/How-many-apartment-buildings-are-there-in-NYC-exclude-condo-co-ops; Tristram Stuart, *Waste: Uncovering the Global Food Scandal* (New York: W. W. Norton, 2009); T. E. Quested et al., "Spaghetti Soup: The Complex World of Food Waste Behaviours," *Resources, Conservation and Recycling* 79 (2013): 43–51, http://doi.org/10.1016/j.resconrec.2013.04.011; Hawken, *Drawdown*.

27 Kimbrell, "Big Lie"; "Agricultural Policies versus Health Policies," Physicians Committee for Responsible Medicine, accessed March 20, 2017,

http://www.pcrm.org/health/reports/agriculture-and-health-policies -ag-versus-health; "Concentration in the Food Industry: The Story behind *The Informant*," CorpWatch: Holding Corporations Accountable, accessed March 20, 2017, http://community.corpwatch.org/adm/pages /food_industry.php.

28 Alan Bjerga and Jeff Wilson, "The Crop Surplus Is Bad News for America's Farms," *Bloomberg Markets*, January 11, 2016, corrected January 13, 2016, accessed March 20, 2017, https://www.bloomberg.com/news /articles/2016-01-12/farm-boom-fizzles-as-u-s-crop-surplus-expands -financial-strain; P. J. Huffstetter, "Fields of Debt: Falling Prices, Borrowing Binge Haunt Midwest 'Go-Go Farmers,'" *Reuters Investigates*, October 31, 2016, accessed March 20, 2017, http://www.reuters.com /investigates/special-report/usa-farm-debt/; David Oppendahl, "The Downturn in Agriculture: Implications for the Midwest and the Future of Farming," *Chicago Fed Letter* 374 (2017), accessed March 20, 2017, https://www.chicagofed.org/publications/chicago-fed-letter/2017/374.

29 Elke Brandes et al., "Subfield Profitability Analysis Reveals an Economic Case for Cropland Diversification," *Environmental Research Letter* 11 (2016): 014009, doi: 10.1088/1748–9326/11/1/014009.

30 Brandon Keim, "Big, Smart and Green: A Revolutionary Vision for Modern Farming," *Wired*, October 19, 2012, accessed March 20, 2017, https://www.wired.com/2012/10/big-smart-green-farming/; Adam S. Davis et al., "Increasing Cropping System Diversity Balances Productivity, Profitability and Environmental Health," *PLoS ONE* 7 (2012): e47149, doi: 10.1371/journal.pone.0047149.

31 J. D. Glover et al., "Increased Food and Ecosystem Security via Perennial Grains," *Science* 328 (2010): 1638–1639, doi: 10.1126/science.1188761; Meghann E. Jarchow et al., "Trade-off among Agronomic, Energetic, and Environmental Performance Characteristics of Corn and Prairie Bioenergy Cropping Systems," *Global Change Biology Bioenergy* 7 (2015): 57–71, doi: 10.1111/gcbb.12096; Sarah M. Hirsh et al., "Diversifying Agricultural Catchments by Incorporating Tallgrass Prairie Buffer Strips," *Ecological Restoration* 31 (2013): 201–211, doi: 10.3368/ er.31.2.201; David van Tassel and Lee DeHaan, "Wild Plants to the Rescue," *American Scientist* 101 (2013): 218, doi: 10.1511/2013.102.218; Larissa Zimberoff, "Superwheat Kernza Could Save Our Soil and Feed Us Well," *Civil Eats*, June 15, 2015, accessed March 20, 2017, http://civileats .com/2015/06/15/superwheat-kernza-could-save-our-soil-and-feed -us-well/.

32 Lee R. DeHaan and David L. Van Tassel, "Useful Insights from Evolutionary Biology for Developing Perennial Grain Crops," *American Journal of Botany* 101 (2014): 1801–1819, doi: 10.3732/ajb.1400084; Lee R. DeHaan et al., "A Pipeline Strategy for Grain Crop Domestication," *Crop Science* 56 (2016): 917–929, doi: 10.2135/cropsci2015.06.0356.

33 J. P. Reganold et al., "Transforming U.S. Agriculture," *Science* 332 (2011): 670–671, doi: 10.1126/science.1202462; Michael Moss, "The Seeds of a New Generation," *New York Times*, February 4, 2014, accessed March 20, 2017, https://www.nytimes.com/2014/02/05/dining/the-seeds-of-a-new -generation.html; Top Producer Editors, "Specialty-Crop: Producers Look Past Corn and Soybeans for Profits," *AG WEB*, October 4, 2016, accessed March 20, 2017, http://www.agweb.com/article/specialty-crop -producers-look-past-corn-and-soybeans-for-profits-naa-top-producer -editors/; "Farmers Look for Profitable Crops in 2016," *PRO AG*, March 8, 2016, accessed March 20, 2017, http://www.proag.com/News /Farmers-Look-for-Profitable-Crops-in-2016-2016-03-08/4437.

34 "Farming Leafy Greens," LGMA, California Leafy Green Products, accessed June 11, 2017, http://www.lgma.ca.gov/about-us/farming-leafy -greens/; Phillip Molnar, "Salad Bowl of the World: At What Cost?" *Monterey Herald*, April 18, 2015, accessed June 11, 2017, http://www .montereyherald.com/article/NF/20150418/NEWS/150419757; Brian Palmer, "The C-Free diet," *Slate*, July 10, 2013, accessed March 4, 2017, http://www.slate.com/articles/health_and_science/explainer/2013/07 /california_grows_all_of_our_fruits_and_vegetables_what_would_we_eat _without.html; Natasha Geiling, "Agriculture Seeks Silicon Valley's Help to Satisfy the World's Demand for Food," *Think Progress*, January 21, 2016, accessed June 11, 2017, https://thinkprogress.org/agriculture-seeks-silicon -valleys-help-to-satisfy-the-world-s-demand-for-food-70d6625d2ed5; Judy Woodruff, "California's 'Salad Bowl' Is Cultivating More Than Crops," *PBS News Hour*, November 30, 2016, accessed June 11, 2017, http:// www.pbs.org/newshour/bb/californias-salad-bowl-cultivating-crops/.

35 Richard Chover, "Organic Food: How Long Does It Take to Ship Produce Once Picked from California to New York?" *Quora*, December 28, 2015, accessed June 11, 2017, https://www.quora.com/Organic-Food -How-long-does-it-take-to-ship-produce-once-picked-from -California-to-New-York; Dana Gunders, "Wasted: How America Is Losing up to 40 Percent of Its Food Farm to Fork to Landfill," NRDC Issue Paper, August 2012, accessed June 11, 2017, https://www.nrdc.org /sites/default/files/wasted-food-IP.pdf.

36 Dickson Despommier, *The Vertical Farm: Feeding the World in the 21st Century* (New York: Picador, 2010); Ian Frazier, "The Vertical Farm: Growing Crops in the City without Soil or Natural Light," *New Yorker*, January 9, 2017, accessed June 11, 2017, http://www.newyorker.com /magazine/2017/01/09/the-vertical-farm; Tammy LaGorce, "How Does This Garden Grow? To the Ceiling," *New York Times*, July 22, 2016, accessed June 11, 2017, https://www.nytimes.com/2016/07/24/nyregion /food-produced-by-the-high-tech-urban-farming-reaches-new-heights .html?_r=0.

37 Rosenberg's talk was the first of five in the session "Green Farming"— see Smithsonian Earth Optimism Summit, Washington, D.C., April 21– 23, 2017, https://earthoptimism.si.edu/live-stream/.

38 Ibid.

39 Ibid.

40 Ibid.; Ramon Taylor, "Profits from Eco-Friendly Vertical Farming Stack Up," *VOA News*, May 20, 2016, accessed June 11, 2017, https://www .voanews.com/a/profits-from-eco-friendly-vertical-farming-stack-up /3337900.html; Chris Clayton, "AeroFarms Plans World's Largest Vertical Farm, $70 M Financing-DTN," *AGFAX*, April 18, 2016, accessed June 11, 2017, http://agfax.com/2016/04/18/aerofarms-pioneering -vertical-farming-aka-indoor-crops/; Caitlin Dewey, "Pioneers of Organic Farming Are Threatening to Leave the Program They Helped Create," *Washington Post*, November 2, 2017; downloaded November 2, 2017, https://www.washingtonpost.com/news/wonk/wp/2017/11/02 /pioneers-of-organic-farming-are-threatening-to-leave-the-program -they-helped-create/?utm_term=.ff505efoeae2.

41 Michael Burns (producer), *The City That Waits to Die*, British Broadcasting Service, 1971. For more about this cult geology film, see Geoblogosphero, "San Francisco, the City That Waits to Die!," posted January 31, 2007, on the *Apparent Dip Blog*, accessed March 20, 2017, http:// apparentdip.blogspot.com/2007/01/san-francisco-city-that-waits-to -die.html; "Rose Canyon Fault," *Wikipedia*, accessed March 17, 2017, https://en.wikipedia.org/wiki/Rose_Canyon_Fault; Robert Monroe, "Fault System off San Diego, Orange, Los Angeles Counties Could Produce a Magnitude 7.3 Quake," *UC San Diego News Center*, March 7, 2017, accessed March 20, 2017, http://ucsdnews.ucsd.edu/pressrelease /fault_system_off_san_diego_orange_los_angeles_counties_could _produce_a_magn; Heather Hope, "Living on the Edge: Erosion Threatens Cliffside Homes in La Jolla," *CBS 8*, September 14, 2015, ac-

cessed March 20, 2017, http://www.cbs8.com/story/30028810/living-on
-the-edge-erosion-threatens-cliffside-homes-in-la-jolla; Nickey Woolf,
"El Niño Storms Cause Cliff to Crumble into Ocean as Houses Teeter
on Edge," *Guardian*, January 26, 2016, accessed March 20, 2017, https://
www.theguardian.com/environment/2016/jan/26/el-nino-storms
-california-houses-crumble-slide-cliff-ocean.

42 Klaus H. Jacob, "Sea Level Rise, Storm Risk, Denial, and the Future of
Coastal Cities," *Bulletin of the Atomic Scientists* 71 (2015): 40–50, doi/
pdf/10.1177/0096340215599777.

43 James E. Neumann et al., "Joint Effects of Storm Surge and Sea-Level
Rise on US Coasts: New Economic Estimates of Impacts, Adaptation, and
Benefits of Mitigation Policy," *Climate Change* 129 (2014): 337–349, doi:
10.1007/s10584-014-1304-z. Given the huge costs enumerated in Neumann's paper, it is important to know that William Nordhaus of Yale
found Emanuel's analysis to be way too conservative based on empirical
historical data. In other words, the reality is likely to be much worse than
Neumann and Emanuel's new analysis suggests. See William D. Nordhaus, "The Economics of Hurricanes and Implications of Global Warming," *Climate Change Economics* 1 (2010): 1–20, doi: 10.1142/S20100078
10000054; William D. Nordhaus, *The Climate Casino: Uncertainty and Economics for a Warming World* (New Haven: Yale University Press, 2013).

44 Richard Parker, "Why Texas Is No Longer Feeling Miraculous," *New
York Times*, September 22, 2017, accessed October 1, 2017, https://www
.nytimes.com/2017/09/22/opinion/sunday/texas-unemployment
-businesses.html?_r=0; Steve Evans, "Hurricane Irma Economic Losses
$58–$83 Billion: Moody's Analytics," *Seeking Alpha*, September 14, 2017,
accessed October 1, 2017, https://seekingalpha.com/article/4106987
-hurricane-irma-economic-losses-58-83-billion-moodys-analytics; Ni-
cole Friedman, "Hurricane Maria Caused as Much as $85 Billion in In-
sured Losses, AIR Worldwide Says," *Wall Street Journal*, updated
September 25, 2017, accessed October 1, 2017, https://www.wsj.com
/articles/hurricane-maria-caused-as-much-as-85-billion-in-insured
-losses-air-worldwide-says-1506371305; Julia Horowitz, "Hurricanes
Irma and Harvey Have Racked up Billions in Damages. Who Pays?"
CNN Money, September 26, 2017, accessed October 1, 2017, http://money
.cnn.com/2017/09/15/news/economy/irma-harvey-damage-who-pays
/index.html.

45 Amanda DeMatto, "Why New York's Sea Level Is Rising Faster Than
the World's," *Popular Mechanics*, July 9, 2012, accessed March 20, 2017,

http://www.popularmechanics.com/science/environment/a7791/why
-new-yorks-sea-level-is-rising-faster-than-the-worlds/; Jeff Goodell,
"Can New York Be Saved in an Era of Global Warming?" *Rolling Stone*,
July 5, 2016, accessed March 20, 2017, http://www.rollingstone.com
/politics/news/can-new-york-be-saved-in-the-era-of-global-warming
-20160705.

46 See the Multihazard Mitigation Council's "Study Documentation," in
*Natural Hazard Mitigation Saves: An Independent Study to Assess the Future
Savings from Mitigation Activities* (Washington, DC: National Institute
of Building Sciences, 2005), vol. 2, accessed March 20, 2017, http://c
.ymcdn.com/sites/www.nibs.org/resource/resmgr/MMC/hms_vol2
_ch1-7.pdf?hhSearchTerms=Natural+and+hazard+and+mitigation; also
National Research Council, *Abrupt Impacts of Climate Change: Anticipating
Surprises* (Washington, DC: National Academies Press, 2010), accessed
March 20, 2017, https://www.nap.edu/read/18373/chapter/1#ii; Jacob,
"Sea Level Rise. "

47 Jeroen C. J. H. Aerts et al., "Evaluating Flood Resilience Strategies for
Coastal Megacities," *Science* 344 (2014): 473–475, doi: 10.1126/science
.1248222; Goodell, "Can New York Be Saved?"; Nathan Kensinger,
"Climate Change in Trump's NYC: How At-Risk Neighborhoods Are
Combating Rising Sea Levels," *Curbed New York*, January 26, 2017, accessed March 20, 2017, http://ny.curbed.com/2017/1/26/14390694
/donald-trump-new-york-climate-change-hurricane-sandy.

48 Goodell, "Can New York Be Saved?"; Emily Badger, "5 Ideas That Could
Have Prevented Flooding in New York," *Atlantic/CityLab*, October 12,
2012, accessed March 20, 2017, http://www.citylab.com/tech/2012/10/5
-ideas-could-have-prevented-flooding-new-york/3754/; Jennifer Peltz,
"NYC Flood Defense Plan Advances, But Completion Years Off," *Phys-
Org*, January 27, 2016, accessed March 20, 2017, https://phys.org/news
/2016-01-nyc-defense-advances-years.html.

49 Alec Appelbaum, "New York's Big Climate Plan Really Does Include
Oysters," *Atlantic/CityLab*, December 15, 2015, accessed March 20, 2017,
http://www.citylab.com/weather/2015/12/new-yorks-big-climate-plan
-really-does-include-oysters/419847/; "Co-op City, Bronx," *Wikipedia*,
accessed February 17, 2017, https://en.wikipedia.org/wiki/Co-op_City,
_Bronx; Erica Pearson, "Troubles for Co-op City, and Middle Class
Housing in NYC," *Gotham Gazette*, October 13, 2003, accessed March 4,
2017, http://www.gothamgazette.com/development/1992-troubles-for
-co-op-city-and-middle-class-housing-in-nyc; Nina Wohl, "Co-op

City: The Dream and the Reality," master's thesis, Columbia University, Graduate School of Arts and Sciences, American Studies Center for the Study of Ethnicity and Race, 2016, http://dx.doi.org/10.7916 /D8MK6CZV.

50 "Rhode Island," *Wikipedia*, accessed March 20, 2017, https://en.wikipedia .org/wiki/Rhode_Island; "1938 New England Hurricane," *Wikipedia*, accessed March 20, 2017, https://en.wikipedia.org/wiki/1938_New_England _hurricane; RI Shoreline Change Special Area Management Plan (Beach SAMP), accessed March 19, 2017, http://www.beachsamp.org/about/; Stephen Long, *Thirty-eight: The Hurricane That Transformed New England* (New Haven: Yale University Press, 2016).

51 Malcolm L. Spaulding et al., "STORMTOOLS: Coastal Environmental Risk Index (CERI)," *Journal of Marine Science and Engineering* 4 (2016): 54, doi: 10.3390/jmse4030054.

52 Malcolm L. Spaulding et al., "Application of State of Art Modeling Techniques to Predict Flooding and Waves for an Exposed Coastal Area," *Journal of Marine Science and Engineering* 5, no. 1 (March 2017), doi:10.3390/jmse5010014.

53 *Restore the Mississippi River Delta*, accessed June 13, 2017, http:// mississippiriverdelta.org/. The coalition includes the Coalition to Restore Coastal Louisiana, the Environmental Defense Fund, the Lake Pontchartrain Basin Foundation, the National Wildlife Federation, and the Audubon Society.

54 Krisha Rao, "Climate Change and Housing: Will a Rising Tide Sink All Homes?" Zillow.com, June 2, 2017, downloaded July 28, 2017, https:// www.zillow.com/research/climate-change-underwater-homes-12890/; Christopher Flavelle, "The Nightmare Scenario for Florida's Homeowners: Demand and Financing Could Collapse before the Sea Consumes a Single House," *Bloomberg*, April 19, 2017, https://www .bloomberg.com/news/features/2017-04-19/the-nightmare-scenario -for-florida-s-coastal-homeowners.

55 Solomon Hsian et al., "Estimating Economic Damage from Climate Change in the United States," *Science* 356, no. 6345 (2017): 1362–1369, http://science.sciencemag.org/content/356/6345/1362; Brad Plumer and Nadja Popovich, "As Climate Changes, Southern States Will Suffer More Than Others," *New York Times*, June 29, 2017, https://www .nytimes.com/interactive/2017/06/ . . . /southern-states-worse-climate -effects.ht; Robinson Meyer, "The American South Will Bear the Worst of Climate Change's Costs," *Atlantic*, June 29, 2017, accessed October 1,

2017, https://www.theatlantic.com/science/archive/2017/06/global
-warming-american-south/532200/.

56 "2003 European Heat Wave," *Wikipedia*, accessed July 28, 2017, https://en
.wikipedia.org/wiki/2003_European_heat_wave. Jeremy and his wife,
Nancy, were there in the old stone farm house they had rented in the
valley of the Célé for seventeen years. Temperatures soared above 100
degrees Fahrenheit for several days and the nights hardly cooled down.
Everyone, including policemen who parked their squad cars next to signs
warning "swimming prohibited," soaked in the rivers to cool off. But
the people in the cities had no such escape, and that's where most of the
deaths occurred.

57 Smithsonian Earth Optimism Summit, https://earthoptimism.si.edu/.

58 Christina Angelides, "The 1990 Clean Air Act Will Save 4.2 Million
Lives by 2020," National Resource Defense Council (NRDC), March 1,
2011, accessed July 28, 2017, https://www.nrdc.org/experts/christina
-angelides/1990-clean-air-act-will-save-42-million-lives-2020; Char-
lotte Tucker, "Number of Lives Saved by U.S. Clean Air Act Continues
to Grow: Opponents Trying to Repeal Protections," *Nation's Health* 41
(May–June 2011): 1–14, http://thenationshealth.aphapublications.org
/content/41/4/1.3.full; Alan H. Lockwood, "How the Clean Air Act Has
Saved $22 Trillion in Health-Care Costs," *Atlantic*, September 7, 2012,
https://www.theatlantic.com/health/archive/2012/09/how-the-clean
-air-act-has-saved-22-trillion-in-health-care-costs/262071/; Joe Romm,
"Trump's War on EPA Regulations Will Kill Jobs and a Lot of People,"
Think Progress, January 25, 2017, accessed July 28, 2017, https://
thinkprogress.org/trumps-war-on-epa-regulations-will-kill-jobs-and
-a-lot-of-people-784727d5f81d.

59 Julie Turkewitz, Henry Fountain, and Hiroko Tabuchi, "New Hazard
in Storm Zone: Chemical Blasts and 'Noxious' Smoke," *New York Times*,
August 31, 2017, accessed October 1, 2017, https://www.nytimes.com
/2017/08/31/us/texas-chemical-plant-explosion-arkema.html?mcubz=0;
Hiroko Tabuchi, "A Sea of Environmental Hazards in Houston's
Floodwaters," *New York Times*, August 31, 2017, accessed October 1,
2017, https://www.nytimes.com/2017/08/31/us/houston-contaminated
-floodwaters.html.

60 "The War of the Currents," Energy.gov, November 18, 2014, accessed
October 6, 2017, https://energy.gov/articles/war-currents-ac-vs-dc
-power; "Who Killed the Electric Car?" *Wikipedia*, accessed June 9, 2017,
https://en.wikipedia.org/wiki/Who_Killed_the_Electric_Car%3F.

61 Michael Moyer, "Climate Change May Mean Slower Winds," *Scientific American*, October 1, 2009, https://www.scientificamerican.com/article /climate-change-may-mean-slower-winds/; "Climate Change Indicators: Weather and Climate," EPA, https://www.epa.gov/climate-indicators /weather-climate; Steve Lang, "NASA Sees Pineapple Express Deliver Heavy Rains, Flooding to California," *NASA Earth*, January 11, 2017, https://www.nasa.gov/feature/goddard/2017/nasa-sees-pineapple -express-deliver-heavy-rains-flooding-to-california; Robert Sweeney, "Warm Spells in Arctic Stunt Crop Yields across US, Study Suggests," *CarbonBrief*, July 10, 2017, https://www.carbonbrief.org/warm-spells-arctic -stunt-crop-yields-across-us-study-suggests.

62 That is what the whole anti-immigration ruckus is all about, the obvious irony being that immigration is and continues to be a huge part of what made America great in the first place. Just look at the diversity of names of the Americans who have won Nobel Prizes, are in the National Academies, fought heroically in our wars, or for better or worse came together for the United States to develop the atomic bomb to hasten the end of World War II.

63 "Love Canal," *Wikipedia*, accessed July 29, 2017, https://en.wikipedia.org /wiki/Love_Canal; John M. Barry, *Rising Tide: The Great Mississippi Flood of 1927 and How It Changed America* (New York: Simon & Schuster, 1997); Anna M. Michalak et al., "Record-Setting Algal Bloom in Lake Erie"; Dan Egan, *The Death and Life of the Great Lakes* (New York: Norton, 2017).

64 Miranda Green, "Governors, Mayors Vow to Continue Green Fight," CNNPolitics.com, June 2, 2017, accessed September 4, 2017, http://www .cnn.com/2017/06/02/politics/governors-mayors-paris-climate/index .html; Ben Bradford, "States And Cities Are Fighting Climate Change, with or without Nations," Capital Public Radio News, April 11, 2017, accessed September 4, 2017, http://www.capradio.org/articles/2017/04 /11/states-and-cities-are-fighting-climate-change,-with-or-without -nations/; UC Newsroom, "University of California Carbon Neutrality Initiative," January 25, 2017, accessed September 4, 2017, https://www .universityofcalifornia.edu/initiative/carbon-neutrality-initiative; Chris Magerian, "Gov. Jerry Brown Signs Climate Change Legislation to Extend California's Cap-and-Trade Program," *Los Angeles Times*, July 25, 2017, accessed September 4, 2017, http://www.latimes.com/politics /essential/la-pol-ca-essential-politics-updates-jerry-brown-climate -change-1500992377-htmlstory.html; Brad Plumer, "How Far Can

California Possibly Go on Climate?" *New York Times*, July 26, 2017, accessed September 4, 2017, https://www.nytimes.com/2017/07/26/climate /california-climate-policy-cap-trade.html?_r=0.

65 Mark Muro and Sifan Liu, "Another Clinton Trump Divide: High-Output America vs. Low-Output America," *Brookings*, November 29, 2016, accessed September 4, 2017, https://www.brookings.edu/blog/the -avenue/2016/11/29/another-clinton-trump-divide-high-output -america-vs-low-output-america/, especially: "The less-than-500 counties that Hillary Clinton carried nationwide encompassed a massive 64 percent of America's economic activity as measured by total output in 2015. By contrast, the more-than-2,600 counties that Donald Trump won generated just 36 percent of the country's output—just a little more than one-third of the nation's economic activity."

66 Dominic Rushe, "Disney, the Gap and Pepsi Urged to Quit US Chamber of Commerce," *Guardian*, April 24, 2017, accessed September 4, 2017, https://www.theguardian.com/business/2017/apr/24/disney-the -gap-and-pepsi-urged-to-quit-us-chamber-of-commerce.

67 Risky Business, *National Report: The Economic Risks of Climate Change in the United States*, accessed October 1, 2017, https://riskybusiness.org /report/national/; Mike Cummings, "Kerry-led Conference Sets Climate-Change Agenda," *Yale News*, September 20, 2017, accessed October 1, 2017, https://news.yale.edu/2017/09/20/kerry-led-conference-sets-climate -change-agenda; John F. Kerry, "There Are No (D)s or (R)s after Storm Names," *Boston Globe*, September 18, 2017, accessed October 1, 2017, https://www.bostonglobe.com/opinion/2017/09/17/climate-change -bipartisan-threat/G8yXxPrAig12SFnzB5pumI/story.html.

68 Thomas F. Steyer, "Why Clean Energy Is the Next Big Business Opportunity," *Risky Business*, December 9, 2016, accessed October 1, 2017, https://riskybusiness.org/2016/12/09/steyer-why-clean-energy-is-the -next-big-business-opportunity/; Michael R. Bloomberg, Henry M. Paulson Jr., and Thomas F. Steyer, "Risky Business: The Bottom Line on Climate Change," *Risky Business*, 2017, accessed October 1, 2017, https:// riskybusiness.org. For a discussion of the amounts of money sitting on the sidelines, and interest potentially to be earned by investing in renewables, see Yale Climate Change Conference, session 2: "The Role of the Private Sector," streamed live on YouTube, September 18, 2017, accessed October 1, 2017, https://www.youtube.com/watch?v=w9xrgE3eafw.

69 Luis Ferré-Sadurní, Lizette Alvarez, and Frances Robles, "Puerto Rico Faces Mountains of Obstacles on the Road to Recovery," *New York Times*,

September 21, 2017, accessed October 1, 2017, https://www.nytimes.com /2017/09/21/us/hurricane-maria-puerto-rico-recovery.html; Frances Robles and Luis Ferré-Sadurní, "Puerto Rico's Agriculture and Farmers Decimated by Maria," *New York Times*, September 24, 2017, accessed October 1, 2017, https://www.nytimes.com/2017/09/24/us/puerto-rico -hurricane-maria-agriculture-.html; Luis Ferré-Sadurní, Frances Robles, and Lizette Alvarez, "'This Is Like in War': A Scramble to Care for Puerto Rico's Sick and Injured," *New York Times*, September 26, 2017, accessed October 1, 2017, https://www.nytimes.com/2017/09/26/us /puerto-rico-hurricane-healthcare-hospitals.html?hp&action =click&pgtype=Homepage&clickSource=story-heading&module =photo-spot-region&region=top-news&WT.nav=top-news&_r=0.

70 George T. Mazuzan, "The National Science Foundation: A Brief History," National Science Foundation, July 15, 1994, accessed March 20. 2017, https://www.nsf.gov/about/history/nsf50/nsf8816.jsp.

INDEX

French Quarter, New Orleans,
72, 76
FreshDirect, 187
Fresh water, 65, 73–75, 88, 92,
94–95, 120, 167
"Freshwater River Diversions for
Marsh Restoration in Louisiana" (Turner), 57
Freudenburg, William, 106
Fruits and vegetables: and agriculture crisis, 149–151, 158; and
agriculture reinvention, 167–168;
and biofuels production, 13–14;
California's production of, 24,
184–185; organic, 26; profitability of, 184–186; waste in
processing of, 186
Fugate, Grover, 194–195

Gainesville, Florida, flooding in,
138
Galeta reef, 60
Galveston, Texas: harbor as
economic growth driver in, 188;
sea level rise impact on, 126
Ganges Delta, 128
Gas. *See* Oil and gas industry
General Motors, 198
Geological Survey (USGS), U.S.,
33, 39, 113, 126
Georgia, drought in, 147
Germany, renewable energy use
in, 175–176
Glaciers, 50, 122–124, 127;
pressure melting of, 122;
rebound of, 127
"A Global Chill in Commodity
Demand Hits America's
Heartland," 205n3

Glyphosate, 27, 34–35, 39–40
GMOs (genetically modified
organisms), 4, 7, 12, 34–35, 156,
181, 184, 187
Go Down Moses (Faulkner), 107
Goldman Sachs, 187
Google, 105, 114, 149, 173,
223n21
Gourmet Sweet (produce), 24–25
Gramling, Robert, 106
Grand Bayou Indians, 46
Grand Bayou Village, 57
"Grand hypothesis," 95–98
Grand Isle, 102
Great Depression, 29, 76, 138
Great Hurricane (Rhode Island,
1938), 194
Great Lakes, 34, 38, 42, 165, 197
Great Miami Hurricane (1926),
138, 141
Great Mississippi Flood (1927),
47, 52, 107, 110, 199
Great South Bay, 131
Greenhouse gas emissions, 18,
20–21, 119, 168, 178, 181
Greenhouse Gas Intensity, 15
Greenland ice sheet, 120–125,
134, 136, 189–190, 196
Groundwater, 66–67, 149–150,
160–161, 163
Gulf of Maine, 56
Gulf of Mexico: dead zone in, 8,
62–68, 73, 104, 108, 113, 161,
165–166; fisheries in, 65;
Mississippi River's impact on,
30, 52, 73, 82; and New
Orleans, 112–113, 115; nutrients
in, 7, 34; and sea level rise, 76,
170–171; and spoil banks, 61–62